D0406976

Rail, Steam, and Speed

Rail, Steam, and Speed

The "Rocket" and the Birth of Steam Locomotion

CHRISTOPHER McGOWAN

COLUMBIA UNIVERSITY PRESS NEW YORK

COLUMBIA UNIVERSITY PRESS
Publishers Since 1893
New York   Chichester, West Sussex

First published in Great Britain by Little, Brown

Library of Congress Cataloging-in-Publication Data
McGowan, Christopher.
Rail, steam, and speed : the"Rocket" and the birth
of steam locomotion / Christopher McGowan.
p. cm.
Includes bibliographical references and index.
ISBN 0-231-13474-6 (cloth:alk.paper)
1. Steam locomotives—England—History—19th century.
2. Railroads—England—History—19th century. I. Title.

TJ603.4.G7M43 2004
625.26'1'094209034—dc22 20040409411

∞

Columbia Universty Press books and printed on
permanent and durable acid-free paper

Printed in the United States of America
c 10 9 8 7 6 5 4 3 2 1

Contents

ACKNOWLEDGEMENTS vii

1 The sport of kings 1
2 Lessons from the past 32
3 The London challenge 62
4 A man of principles 80
5 Up from the mine 100
6 Famous son of a famous father 135
7 *Rocket* on trial 157
8 The people's choice 191
9 The dark horse 201
10 Winning day 216
11 Triumph and tragedy 231
12 Boom and bust 262
13 Beginnings and endings 270

NOTES 317
INDEX 354
CREDITS 380

Acknowledgements

The greatest reward in writing this book has been meeting a cadre of such knowledgeable enthusiasts, from industrial archaeologists and railway historians to engine crews and the builders of the replica engines they drive. I have an unredeemable debt of gratitude to each one of you for your generosity and patience in giving me a better understanding of your fields of expertise. You have answered my questions, explained your engines, discussed your research, provided reprints, preprints, unpublished manuscripts and archival material. I give my sincerest thanks to each one of you. With apologies for any omissions, I thank John Allen, Philip Atkins, Paul Belford, Tony Burton, Colin Divall, Francis Evans, Patrick Greene, Andy Guy, Richard Hayman, David Heaton, Richard Hills, Dieter W. Hopkin, Lars Olov Karlsson, Richard Lamb, Michael Lewis, John Liffen, John Selway, Peter Stokes, Jennifer Tann and Roger Waldron.

Michael Bailey, an eternal font of railway history, was kindness personified with his generous and indulgent help. He replied to numerous e-mails, assisted with literature, suggested names I should contact, and gave freely of his extensive knowledge of the subject. Our long discussions on *Rocket* and the Stephensons have been invaluable, and his meticulously careful reading of an earlier version of the entire manuscript has greatly improved this work. Richard Gibbon similarly read the entire manuscript, pointing out the errors of my ways, and shared his current and unpublished joint research, with Richard Lamb, on *Novelty*'s air-blower. During my many visits to the National Railway Museum (NRM), York, he discussed engines and engineering, clambered beneath the replica of *Rocket* to demonstrate the reversing mechanism, and facilitated my interactions with the crew when the engine was in steam. Jim Rees shared his knowledge and enthusiasm for early locomotives and the men who built them, from his unique perspective of having built and driven replica locomotives himself. I have gained so much from our discussions. John Glithero has also been more than generous in discussing early locomotives, particularly *Rocket*, and its engineering history, and in reading parts of the manuscript. Dave Burrows and Ray Towell, the replica *Rocket*'s crew, spent much time with me aboard their engine while in steam, answering my many questions and giving me insights into what is involved in its operation. Dave Burrows also took time from a busy schedule to discuss *Rocket* and the Llangollen re-enactment, and to correspond with me later. The crew of *San Pareil*'s replica, David Heaton and Tony Newton, spent much time with me while firing and driving their engine. The information they shared on the operation of the locomotive was invaluable. Sheila Bye spent one Sunday morning with me at the Middleton Railway discussing Blenkinsop's locomotive. She also freely shared information and archival material. Jane Hackworth-Young, the great-great-granddaughter of railway pioneer Timothy Hackworth, provided me with insights into her celebrated ancestor and his family, and gave me a copy of his portrait. She also provided warm hospitality during a recent visit to her pastoral corner of England, gave a guided tour of historically important sites in and around Shildon, and introduced me to the crew of *Sans Pareil*'s replica. Her cousin, Ulick Loring, also shared his knowledge of Timothy Hackworth.

Much of the research for this book was conducted in libraries, archives and museums in various parts of England. Space does not allow me to thank each of these institutions for their generous help, but I would like to acknowledge the Public Record Office and its user-friendly staff: I am as impressed by the extent of the collection as I am with the speed with which the material is located. Keith Moore of the Institution of Mechanical Engineers helped with archival material during my visits, and provided valuable assistance and information afterwards. His assistant, Sarah Vinsen, was also exceedingly helpful. So too were Rheanna Sullivan, formerly of the Science and Society Picture Library, and Mike Chrimes of the Institution of Civil Engineers. Camilla Harrison, NRM, provided valuable logistic support.

Andrew Forester, my long-time friend from undergraduate days and a stickler for the proper use of English, read the entire manuscript most critically, generating reams of comments. His correcting of so many of my transgressions and embarrassing gaffs has improved the manuscript considerably.

This book began as an idea for a complete history of steam power, from which my agent, Jill Grinberg, wisely steered me away. During a discussion with Peter Robinson, our affiliate agent in London, he suggested focussing on Rainhill and locomotives, which was such sound advice. Jill, as always, has given her support and encouragement throughout the entire project.

Working with the professionals at Time Warner Books and Columbia University Press has been a great joy. Tim Whiting and Robin Smith, my editors in the UK and the U.S., have been equally supportive and enthusiastic. Their careful reading of the drafts and clear guidance in helping me make significant organisational changes were critical to the project. Michael Haskell has been most helpful as production editor, and Rachael Ludbrook and Linda Secondari were instrumental in the design of the gorgeous cover.

My sincere thanks to you all.

Last, but by no means least, I wish to thank Liz, my stoic wife, who suffered through my long absences from the real world as I shut myself away in my office – thoroughly boring and antisocial – working from pre-dawn to post-dusk. She also lent her professional assistance, critiquing, proofing, and checking. And the support she gave during the terminal nightmare with the nameless software that refused to index properly is beyond belief. I cannot find words even approaching adequacy to express my thanks for all of this, and for so very much more. I count my blessings.

To Emma, Carter, and Miles, with love.

The Rainhill Trials

CHAPTER 1

The sport of kings

On the morning of Tuesday 6 October 1829, a huge crowd, estimated at between ten and fifteen thousand, gathered at a temporary meeting ground in the north of England, some ten miles from Liverpool. They came from across the land and across the sea, converging on the tiny hamlet of Rainhill, a name that would become synonymous with the event they were to witness. Carriages of every description lined the perimeter of the grounds, but many spectators had arrived on foot, some travelling many miles from the surrounding countryside, just to be there. Pedestrian traffic competed with horse-drawn vehicles, all but clogging the roads leading to the grounds as more people flooded in.

A 'commodious tent' had been erected for the accommodation of the ladies, but many preferred to mill around the grounds, or to enjoy the spectacle from their open carriages. Union Jacks fluttered, a band played and hawkers sold their wares. It might have been a day at the races – the St Leger or the Derby. But this crowd, which included a large contingent of engineers and men of science, had not come to indulge in the sport of kings. They had come to witness the most remarkable event of the Industrial Age: a competition between railway locomotives to see if any one of them was fast enough and powerful enough, and sufficiently reliable and economical, for regular railway service.

Rainhill would be a defining moment in railway history, but the reason why the trials were so pivotal may not have been apparent to many of the spectators. Steam locomotives, after all, had been in existence for a quarter of a century, and the first public railway, the Stockton & Darlington, had been in operation for over four years. During that railway's triumphant opening, George Stephenson's engine, *Locomotion*, had hauled a special train crammed with over 600 passengers. The 400 ft-long train, with people clinging to the outside like bees, reached the amazing speed of 12 mph. This was a spectacular achievement, and the future of the railways seemed assured. But steam locomotives did not live up to their vaunted expectations.

First and foremost, locomotives were notoriously unreliable, and spent a good part of their time in the engine shed. They were sometimes reluctant to start, and might run so low on steam that they had to be coaxed along by their ambulatory crews. They were smoky, and were often accused of terrifying livestock and posing fire hazards with the sparks they belched from their tall chimneys. But the most devastating damage they caused was when their boilers blew up, which happened from time to time. On 1 July 1828 the boiler of *Locomotion* itself exploded while she was taking on water. John Cree, the driver, was killed outright, while Edward Turnbull, the water pumper, was maimed for life. It was said that his face was permanently speckled in the blast, like a dalmatian dog's.

The Stockton & Darlington was a mixed railway, steam

locomotives and horse-drawn wagons sharing the same thirty-one-mile stretch of single track, with sidings every quarter-mile to allow for passing. The railway was built primarily for carrying coal from the rich Durham coalfields, in the west, to the shipping wharves built along the River Tees at Stockton, in the east. Passengers formed only a small portion of the traffic, and they were conveyed only in horse-drawn vehicles. There were two steep inclines along the way, and stationary steam engines, sometimes referred to as 'fixed engines', were used to haul the wagons to the top of the grades, using towing ropes as thick as a man's wrist. The use of fixed engines was a cheap solution to the problem of constructing rail lines in the rolling countryside of the mineral-rich northern counties. Indeed, many people thought rope haulage was superior on the flat too, because of the greater reliability of fixed engines over the more temperamental steam locomotives.

A horse-drawn passenger wagon, used on the Stockton & Darlington Railway.

When the Stockton & Darlington Railway opened, late in September 1825, the economic effect was immediate. Coal prices in Stockton fell from 18*s* to 12*s* a ton, eventually dropping to 8*s* 6*d*. The value of shares soared as people clamoured to invest in the railway. During the inaugural banquet at Stockton's town hall, one gentleman offered to purchase any number of shares at a premium of £20 (£1,000 today). Another said he would offer £30 above the offering price, and soon £40 was being freely offered. But there were no takers because nobody wanted to part with their valuable shares. The initial success of the railway spawned a rash of proposals for others. By far the most important of these was a rail link between the town of Manchester and the port of Liverpool.

Manchester enjoyed the distinction of being the world's first industrial town. Its rapid growth, from a quiet rural village beside the River Irwell to a bustling metropolis, is largely attributable to its geographical location on a navigable river, close to coal mines and other mineral resources. Manchester's booming economy was centred upon the cotton industry, the nearby seaport of Liverpool being a major gateway for the import of goods from the Americas. The city became the entrepreneurial heart of the nation, and that nation was leading the world in the pell-mell dash towards industrialisation. Manchester, then, was a befitting cradle for the most ambitious engineering project of the times.

The directors of the Liverpool & Manchester Railway Company were committed to using steam power: there would be no horse-drawn traffic on this railway. But they were divided on whether to use locomotives or fixed engines and rope haulage. To help them reach a decision they hired two consultants, James Walker and John Urpeth

Rastrick, both of whom were highly respected engineers. Walker was a civil engineer, who later became President of the Institution of Civil Engineers. Rastrick, who had built several locomotives, including *Agenoria* – now on display at the National Railway Museum, York – was much sought after as an adviser, surveyor and engineer of railways. Messrs Rastrick and Walker's mandate was to visit a number of working railways – all in the mining north – and report back on the comparative merits of the two systems.

The consultants presented their findings in two separate reports, but, as Walker pointed out in his paper, they reached the same conclusion, and the differences in their respective calculations were only marginal. With so few railways in operation at that time, their information for comparing the costs of the two systems was quite limited, and drew heavily upon the Stockton & Darlington Railway. According to Walker's estimates, the capital costs came out in favour of locomotives, by a margin of £9,898 6*s* 9*d*. However, the running costs favoured fixed engines, by a margin of £10,176 8*s* 7*d*, and on the strength of this the recommendation was made in favour of stationary engines.

George Stephenson (1781–1848), the company's forty-eight-year-old Engineer-in-Chief, was incensed by the recommendations. He had built his first locomotive fifteen years earlier, and was a staunch advocate of steam locomotion. If the company adopted fixed engines it would put a blight on the locomotive, and Stephenson, a fiercely proud man, would take that as a great personal defeat. He also stood to lose considerable revenue from the locomotive factory he owned with his son, Robert, and other partners. Not that he was a mercenary man: during his early days with the Stockton & Darlington Railway he had recommended their using wrought-iron rails rather than the inferior cast iron ones he had a financial interest in selling them.

The bluff northcountryman was not about to see his dream of a nation connected by railways evaporate before his eyes. Defeat was simply not part of the limited vocabulary of a man who had worked his way up from operating a colliery engine to becoming the premier railway engineer in the land. He therefore sat down and wrote a very restrained critique of the consultants' reports. As he still found writing a challenge – he never went to school and learned to read and write only in his late teens – he would have sought help with this task, probably from his son Robert.

In his rebuttal, Stephenson pointed out that the Stockton & Darlington Railway traversed two steep hills that necessitated rope haulage. It was therefore an inappropriate model for the new railway, which was essentially flat. Stephenson also disagreed with Walker's estimate of the number of locomotives needed to carry the anticipated traffic. Based on his comparisons with the Stockton & Darlington Railway, Walker had arrived at the staggering figure of 102 locomotives. But what his calculations failed to take into account was that the Stockton & Darlington was essentially a one-way railway: coal was hauled from the collieries in the west to the shipping wharves in the east, and the wagons returned empty. There would be two-way traffic between Liverpool & Manchester, with about the same tonnage of goods being hauled in either direction. Stephenson's estimate of the number of locomotives was less than half that of Walker's.

Stephenson's 'observations' were read to the board on 13 April 1829, during one of their regular weekly meetings. But the discussion on the respective merits of locomotives and fixed engines was postponed until the following week.

Just before the meeting adjourned, the directors' attention was drawn to the case of Sarah Smith. Her husband had been killed in an accident at Olive Mount, just outside Liverpool, the site of the deepest railway cutting of the entire line. The directors granted the widow £15 compensation (less than £900 in today's terms). Working-class lives were not valued highly in those days.

At the next board meeting, the directors decided to adopt an idea contained in Walker's report. They would offer a prize, settled at £500 (about £30,000), 'for a Locomotive Engine which shall be a decided improvement on those now in use, as respects the consumption of smoke, increased speed, adequate power, and moderate weight . . .'. Although the prize was substantial, the opportunity of supplying locomotives to the company was a far greater incentive. If that did not inspire the muse of invention, the locomotive had a doubtful future.

The draft of the stipulations and conditions of the contest, together with the wording of the advertisement that would appear in the press, was submitted to the directors for their approval at the 27 April board meeting. The public announcement of the trials was made a day or so later, giving prospective competitors only a little over five months to design and build their locomotives. Stephenson would have had the advantage of about an extra week. Also, as engineer-in-chief, he would have had a large part in drawing up the list of stipulations.

In spite of the remarkably short notice, the directors received many enquiries about the competition, but some were less serious than others. This was an age of technological wonders, both real and imagined, and all manner of wild and fanciful ideas were submitted, as described by Henry Booth (1789–1869), the company's treasurer:

> The friction of the carriages was to be reduced so low that a silk thread would draw them, and the power to be applied so vast as to rend a cable asunder. Hydrogen gas and high-pressure steam – columns of water and columns of mercury – a hundred atmospheres and a perfect vacuum – machines working in a circle without fire or steam, generating power at one end of the process and giving it out at the other . . . wheels within wheels, to multiply speed without diminishing power – with every complication of balancing and counterbalancing forces, to . . . perpetual motion – every scheme which the restless ingenuity or prolific imagination of man could devise was liberally offered to the company . . .'

But there were also a number of serious contenders, and it seemed there would be a strong field of starters. As it happened, only five competitors showed up on the first day of the trials, time having run out for the other hopefuls.

The directors, each wearing a white ribbon in his buttonhole, arrived at the course shortly after ten o'clock on a rather chilly morning. The entire line between Liverpool and Manchester was nearing completion, and the two-mile section where the trials would take place had been chosen because it was both straight and level. Unlike the Stockton & Darlington Railway this line was double, but only one track would be used at any one time during the trials. The directors had arranged for over two hundred men to act as special constables to keep the crowds off the line. But their attempts were futile and peopled walked across the tracks as they pleased. Hopefully there would be no mishaps when the locomotives began steaming up and down.

Few of the people milling around the grounds on that

early autumnal Tuesday had ever seen a steam locomotive before, though probably everyone had heard or read about them. The newspapers of the day were full of accounts of the contest, and many of the local and national newspapers had sent reporters to cover the event. The trials were expected to last for three days, ending on the Friday. The official listing of the five contestants, and the order in which they were to be tested, was widely publicised for the benefit of the spectators, and the following appeared in *The Leeds Mercury*:

> No. 1 – Messrs Braithwaite and Erickson [*sic*], of London: 'The Novelty'; weight 3 tons, 15 cwt [hundredweight – 112 lbs].
> 2. – Mr [H]ackworth, of Darlington; 'The Sans Pareil'; weight 4 tons, 8 cwt, 2 qrs [quarter – 28 lbs].
> 3. – Mr Robert Stephenson, Newcastle-upon-Tyne; 'The Rocket'; weight 4 tons, 3 cwt.
> 4. – Mr Brandreth of Liverpool; 'The Cycloped'; weight 3 tons . . .
> Mr Burstall, of Edinburgh, did not bring his carriage out, in consequence of an accident on its road from Liverpool to the course. The damage will, however, be repaired and the machine will, it is expected, be ready by tomorrow.

Fifty-three-year-old Timothy Burstall (1776–1860), the oldest of the contestants, was a Scottish engineer with some experience in steam locomotion. Five years earlier he had built a steam-powered road carriage, and this was followed by a second one. He had shipped his locomotive, *Perseverance*, by sea, consigning it to the port of Liverpool, just a few miles away from the grounds. This was the only way of transporting heavy goods for any distance in those days because the roads were generally so poor. The locomotive was then transferred to a road carriage, which took it to the Liverpool & Manchester Railway Company's Millfield Yard. It was there that the locomotives were to be assembled, prior to being transported to the test track at Rainhill. *Perseverance* was one of the first locomotives to arrive, and had been there, along with *Rocket*, for just over a week. But when *Perseverance* was in transit from the yard disaster struck. A retaining chain became loose, and the locomotive 'had fallen to the ground, by which one of the Cranks was injured, and a pipe broken'. Although the damage did not sound extensive, the repairs would keep Burstall busy until close to the end of the competition.

The directors had appointed three judges to oversee the competition and set the rules: John Urpeth Rastrick, whom we have met, Nicholas Wood and John Kennedy. Wood was a respected engineer, and the author of an erudite textbook on railways. He had met George Stephenson when he was sixteen, having gone to work at the same colliery. Stephenson was fourteen years his senior, and the two seem to have got on well together and had kept on good terms ever since. Kennedy was a wealthy cotton processor from Manchester, who had made his fortune by improving textile machinery.

Some of the spectators would have wanted to take a closer look at the machines before they were demonstrated before the crowd – each one as different from the next as the men who had designed them. But *Novelty*, the first on the list, was the most radically different of them all. Perhaps it was no coincidence that her builders, thirty-two-year-old John Braithwaite (1797–1870) and twenty-six-year-old John

Braithwaite and Ericsson's Novelty.

Ericsson (1803–89) were so young. Those who had seen illustrations of contemporary locomotives would have expected to see the traditional barrel-shaped boiler with its tall elephant-trunk chimney. But *Novelty* had neither of these features. Looking more like some kind of mechanised tea trolley, she was essentially a flat-bed platform, mounted on four large and somewhat spindly looking carriage wheels. Her tea-parlour appearance was enhanced by the presence of a chest-high copper vessel at the front that looked like a coffee urn. This was the steam chamber. The copper box at the back, which might have contained coffee beans, was part of the air compressing system. A short vertical tube at the rear served as the chimney. The platform, which was surrounded by a safety rail, was remarkably uncluttered, providing ample standing room for the two operators, as well as space for a basket of coke. The paired cylinders were mounted vertically, close to the rear wheels; the remainder of the working parts, together with the boiler and water tank, were located below the platform. The blue paintwork complemented the burnished glow of the copper, presenting a most pleasing effect. 'The great lightness of this engine . . .,' wrote a correspondent for the popular *Mechanics' Magazine*, 'its compactness, and its beautiful workmanship, excited universal admiration . . .'. Some spectators may have had difficulty visualising how *Novelty* could generate sufficient power to propel it along at any speed. Part of its apparent impotence stemmed from its having so much of its machinery hidden from view beneath the platform. Also, a cold and inert locomotive was a very different proposition from one that was in steam.

Timothy Hackworth's entry, *Sans Pareil*, was the heaviest locomotive in the competition – solid and seemingly dependable, like the God-fearing man who had built it. As it happened, Hackworth (1786–1850) was born in the same tiny Northumbrian village as George Stephenson, but some five years later. A greater contrast to *Novelty*, the lightest competitor, could scarcely have been found. Everything about *Sans Pareil* looked massive, from her squat boiler to her substantial chimney. Some contemporary illustrations depicted her tender – the wagon that carried the water and fuel – as being coupled behind the engine, as it was in most locomotives. But *Sans Pareil* pushed her tender from the front, a consequence of having a return-flue boiler.

A locomotive's flue was a wrought-iron tube, some 2 ft in diameter, which ran through the boiler barrel, connecting furnace with chimney. Its function was to heat the water: as the hot gases passed through the flue tube, *en route* to the chimney, much of the heat was transferred to the surrounding water in the boiler. The simplest arrangement was to

Hackworth's Sans Pareil *pushed her tender, a consequence of the return-flue boiler.*

have a straight-through flue, which passed in a straight line between furnace and chimney. This was the kind George Stephenson had built for the Stockton & Darlington Railway. The return-flue boiler was more complex in having a U-shaped flue tube that looped back upon itself just before reaching the far end of the boiler. The chimney was therefore at the same end of the boiler as the furnace, and this was at the front of the locomotive. Return-flue boilers were much more difficult to build than the straight-through ones, but had the singular advantage of increasing the total surface area available for transferring heat to the water, thereby increasing the rate of steam production. When Hackworth was a young man he helped build a locomotive with a return-flue boiler, some time between 1811 and 1812, and had favoured them ever since.

Like the other competitors, *Sans Pareil* had paired pistons, and these were placed vertically, directly above the rear wheels. Each piston was coupled to its respective wheel by a connecting rod. A heavy coupling rod linked the front and rear wheels on either side, giving four-wheel drive. She looked a powerful machine, for which the appellation 'iron horse' would have been eminently suitable. But she was obviously a draught horse rather than a thoroughbred. *Sans Pareil*'s racing colours were green, yellow and black.

A curious spectator would have had a myriad of questions for her creator. What were all those complicated levers and rods at the rear end, near the pistons? Why were they connected to the back axle? Hackworth may have replied that it was all part of the reversing mechanism, to change the direction of travel of the engine. He might even have attempted explaining how it worked, but it is doubtful he could have spared the time. He had not had the opportunity of testing *Sans Pareil* before shipping her to Rainhill, and still had much to do to prepare her for the trials.

Spectators who had not read the racing card carefully would have been forgiven for seeing the names Stephenson and *Rocket*, and unconsciously linking the two with George. George Stephenson had not quite become a household name, but anyone interested in steam locomotion would certainly have heard of him. It would therefore have been perfectly natural to assume that *Rocket* had been built by George Stephenson. Ask anyone today and they will probably say that George Stephenson not only built *Rocket*, but also invented the steam locomotive, though neither assertion is true. Dr Dionysius Lardner, prolific author, lecturer, gadfly and sometimes windbag (he once predicted that fast rail travel was impossible because passengers would become asphyxiated) bestowed the title of 'Father of the Locomotive' on George Stephenson in 1836, during a public lecture on steam engines. Samuel Smiles, Stephenson's biographer, perpetuated the myth. In comparing George Stephenson's achievements with those of James Watt, he wrote that while 'Both have been described as the improvers of their respective machines . . . they are rather entitled to be described as their Inventors.' The undeniable truth is that the first steam locomotive was built by Richard Trevithick (1771–1833). Although Trevithick played no direct part in the trials – as far as we know he was not even there – he is the unsung hero of the story that culminated at Rainhill. The remarkable life of this charismatic Cornishman is entwined throughout this narrative.

Rocket was built by Robert Stephenson (1803–59), funding being provided by his father and by Henry Booth. Robert, who would be celebrating his twenty-sixth birthday a few days after the trials, was the youngest contestant at Rainhill. His dashing good looks would also have won him the approbation of the young ladies as the most handsome one too. But he was no longer an eligible bachelor, having married just three months before. He had the same practical way with machines as his father, but, in contrast, he was always willing to seek the advice and help of others. He was also modest and freely acknowledged any help he received, characteristically downplaying his own part along the way. Such humility had certainly not been inherited from George Stephenson. Nor had Robert inherited his father's irrational jealousy that so often led him to make enemies of those he perceived as rivals in the field. But Robert did have his father's dogged determination to succeed against all odds, and he did occasionally display a rare flash of his father's arrogance. Conceit, though, was alien to the natural modesty

Robert Stephenson, the youngest of the Rainhill contestants.

of the diffident young man who 'charmed all who came in contact with him . . .'.

Compared with Hackworth's *Sans Pareil*, *Rocket* was a horse of an entirely different colour. Her pistons were mounted near the back, but were inclined obliquely, to connect with the front wheels. These front driving wheels were much larger than the rear wheels, whereas in *Sans Pareil* all four wheels were the same size. There were no coupling rods between the wheels, so *Rocket* had two-wheel drive. And, whereas *Sans Pareil*'s furnace was little more than an extension of the flue tube, *Rocket* had a discrete firebox, partly encased by a copper water jacket. The gleaming copper complemented her striking black, yellow and white livery. Both machines had a tall chimney, though *Rocket's* was braced on either side by a tie rod, for stability when the engine was in motion.

The fortunate few who got near enough to *Rocket* for a

Stephenson's Rocket.

close-up view must have been thrilled beyond measure. Just earlier that morning she had conveyed the directors to the grounds, and was still in steam. Here was a living, breathing locomotive, all heat and steam and cinders, dripping water and smelling of smoke and hot metal. But elation may have turned to alarm when vented steam suddenly hissed from the safety valve, invoking fears of bursting boilers.

Unlike the other entries, *Rocket* had been designed and constructed in a purpose-built locomotive factory, with the best facilities and most experienced work force in the land. As a consequence she was essentially finished several weeks before the start of the trials, allowing time for a day's testing on the track at Killingworth Colliery before being shipped to Rainhill. *Novelty*, in contrast, could not have been track-tested even if she had been ready on time, simply because there were no railway tracks in London – all the railways were north of the capital. As *Rocket* was the only locomotive ready to proceed, the directors decided she should entertain the crowd with a demonstration of her abilities. To this end a number of wagons were hitched to the engine and loaded with stones, equal to three times her estimated weight, a load the judges deemed appropriate for the trial. A section of the line was then staked out for the demonstration. It was a little over one and a half miles long, and was centred upon the marquee that had been erected for the ladies. All was ready.

Operating a steam locomotive required two men, a fireman and driver. *Rocket*'s fireman and driver shared a narrow platform at the rear of the engine. The fireman's job was to maintain steam pressure by shovelling fuel from the tender to the furnace, and to pump water into the boiler from the tender. The earlier locomotives ran on coal, but one of the stipulations of the Act of Parliament authorising the Liverpool & Manchester Railway was that the locomotive should 'consume its own Smoke . . .'. While some coals produced very little smoke, coke, made by heating coal in a retort to drive off the gas and tar, was the only truly smokeless fuel. Consequently all the contestants were required to fuel their locomotives with coke.

Locomotive drivers of those days were highly skilled, and very well paid for their services. They had to be proficient at their job because their charges were so difficult to handle. For example, locomotives had no brakes, and the only way to slow them down was to use the pistons to oppose forward motion. Understanding how this worked requires knowledge of how steam was supplied to the pistons.

Steam was introduced sequentially, to either end of the cylinder, via a valve. Consequently, every stroke of the piston was a power stroke. The distance the piston moved during each stroke was carefully matched to the diameter of the driving wheel, and to the attachment point of the piston coupling. This was so that each traverse of the piston advanced the wheel one half-turn. The next stroke of the piston advanced the wheel another half-turn, thereby completing one revolution. The two opposing pistons had to be out of phase with one another, something like the two pedals of a bicycle, so that motion could be sustained.

The valve supplying steam to the cylinder had to be synchronised with the revolutions of the wheel – the faster the locomotive moved the more rapidly the steam had to be alternated between the two ends of the cylinder. This was achieved by coupling each valve, by a series of linking rods, to an eccentric – a type of cam – attached to one of the wheel axles.

The eccentric is a device, conceived long before locomotives, for converting circular into reciprocating motion. It is remarkably simple, being a disc attached to a rotating shaft by a hole set off-centre, with some tracking device with which it makes contact. In the early locomotives the eccentric had the form of an iron disc surrounded by a circular collar, called an eccentric strap. The latter was connected to a rod which, through linkages, provided reciprocating motion to the valves.

To slow a locomotive the driver had to reverse the sequence of injecting steam into the two ends of the cylinder. In this way a stroke that was pushing the wheel forward became one that pushed it back. This was achieved by reversing the valve sequence, and was equivalent to putting a car into reverse gear, but the results were quite unspectacular, causing the locomotive to slow to a gentle halt. To do this the driver first had to slow the locomotive by cutting off the steam to the cylinders, until it was travelling at less than about 8 mph. Next, he had to disengage the linkage between the forward-drive mechanism and the valves, and then engage the reverse-drive mechanism. Steam was then readmitted to the cylinders, forcing the locomotive to slow to a stop. If the steam was not shut off at this point the engine would continue in reverse, so this was also the procedure for changing directions.

It is not clear who drove *Rocket* at Rainhill that day. It may have been her regular driver, a man named Wakefield. But it is as likely to have been George Stephenson, who was never bashful when it came to standing in the limelight. Indeed, with an enthusiastic crowd of at least ten thousand spectators, it is difficult to imagine him not taking centre stage. Regardless of whose hands were on the controls, the driver swung the lever of the regulator valve to the fully open position, at the centre of its travel, releasing the maximum flow of steam into the paired copper pipes supplying the cylinders. *Rocket* slowly inched forward, trailing wisps of steam and droplets of water. Steam locomotives do not accelerate rapidly, and as she majestically gained speed, she began making a sound characteristic of her kind: a low soporific soughing, like the laboured breathing of an asthmatic uncle. And as she gathered speed the soughing quickened into the familiar chuffing of a steam locomotive.

The sight of the black-and-yellow locomotive thundering down the line with its heavily laden wagons thrilled the crowd, and there was a 'simultaneous burst of applause'. Hats waved, voices cheered, handkerchiefs fluttered in the breeze. In no time at all she was a mere smudge in the distance. And then she was back again, charging down the track like some mechanical creature, eating fire and snorting steam. *Rocket* continued racing up and down for about sixty minutes, reaching speeds of just over 13 mph. The loaded wagons were then detached, and *Rocket*, released of her burden, went 'shooting past the spectators with amazing velocity', reaching a staggering 24 mph. This was probably the fastest a vehicle had ever travelled before, and the crowd showed their enthusiastic appreciation.

George Stephenson appears to have been quite relaxed and confident during *Rocket*'s debut, inviting several people to go for a ride with him at the end of the day. Recounting the event more than half a century later, one of the witnesses wrote:

> In the evening after the first day's trial, Mr. Stephenson said to my father, 'Now Stannard, let us have a trip to ourselves;' whereupon he got up, followed by Wakefield, the driver, my father, Mr. Booth, Mr. Moss, and myself.

> There was not much room, and Mr. Stephenson remarked, 'Put your boy up on the tub, Stannard; he'll be more out of harm's way there;' and there indeed I sat during my first ride on the *Rocket*. As the butt had been newly painted only the day before, I stuck on very literally while we ran some four miles out and back.

As *Rocket* performed before the enthusiastic crowds that first morning the most observant spectator might have noticed she rolled slightly, from side to side, with each thrust of the pistons. This was because of the inclination of the cylinders. The fault was eventually corrected, some time after the trials, by repositioning them nearly horizontally. The *Mechanics' Magazine* was quite critical of *Rocket* in its first report of the trials: 'The faults most perceptible in this engine were a great inequality in its velocity,' the article reported, 'and in its very partial fulfilment of the condition that it should "effectually consume its own smoke".' It was subsequently discovered that some coal had inadvertently been mixed with the coke on the first day, causing *Rocket* to smoke; in all likelihood coal had been used to start the fire, coke being notoriously difficult to light with kindling. Although the magazine pointed out this error in a later issue, its reporting continued to be pointedly biased against *Rocket*. However, this was primarily an attack on George Stephenson, and for reasons that will become apparent later.

At this point in the proceedings the thoughts of many spectators must have been turning from steam to stomachs as they contemplated what to do for lunch. Refreshment arrangements were sadly wanting, and little was offered the hungry crowds besides bread and cheese and beer. The few taverns in the vicinity were packed to capacity. The

Large crowds attended the first day of the Rainhill trials, as depicted in this somewhat romanticised illustration.

landlady of the Railroad Tavern at Kenrick's Cross had 'very prudently reserved one room for the accommodation of the better class of visitors whom she treated with boiled beef and roasted mutton for the moderate charge of 3*s* a head . . .' (about £9).

The second locomotive to be exhibited was the *Novelty*, often referred to as 'the London engine'. Her creators, John Braithwaite and John Ericsson, were at the controls for her inaugural run. Ericsson, a former officer in the Swedish army, was a junior partner in Braithwaite's engineering works in London, but he was probably the intellectual leader of their locomotive enterprise. He was arguably the brightest contestant at Rainhill.

For reasons unknown, Ericsson did not learn of the trials until the end of July, giving him and Braithwaite only seven weeks, rather than five months, to build their locomotive. Inevitably, it arrived at the grounds untried and unready. The judges, recognising their problem, decided *Novelty* did not have to haul a load during her demonstration. Such magnanimity may seem uncharacteristic of judges, but this was to be a trial, and if they disqualified every locomotive that was unready to compete on the first day, *Rocket* would win by default, which would serve nobody's interests. The judges informed Messrs Braithwaite and Ericsson that they could merely run up and down the line unencumbered. This, incidentally, saved the judges from having to deal with the thorny problem of deciding what her assigned load should be. Unlike *Rocket* and *Sans Pareil*, which hauled their fuel and water in a tender, *Novelty* carried hers aboard, and some allowance would have to be made for this. But that could wait until the morrow, when *Novelty* was scheduled to undergo her trial – Braithwaite and Ericsson had assured the judges they would be ready in time.

Novelty was lighter than *Rocket.* Some contemporary accounts said she weighed only half as much, but that was an exaggeration because she was about three-quarters as heavy. She looked much racier than *Rocket*, and there was probably much speculation among the spectators as to whether she would be any faster. Bets were likely wagered on the outcome of the afternoon's entertainment.

As *Novelty* was only giving a demonstration run the judges dispensed with any formalities, just as they had for *Rocket.* Once Messrs Braithwaite and Ericsson were satisfied their machine was ready, they were free to proceed. The frock-coated gentlemen did their final checks of their machine, the word was given, and they were away.

> Almost at once, it darted off at the amazing velocity of twenty-eight miles an hour, and it actually did one mile in the incredibly short space of one minute and 53 seconds! Neither did we observe any appreciable falling off in the rate of speed; it was uniform, steady, and

> continuous. Had the railway been completed, the engine would, at this rate, have gone nearly the whole way from Liverpool to Manchester within the hour; and Mr. Braithwaite has, indeed, publicly offered to stake a thousand pounds, that as soon as the road is opened, he will perform the entire distance in that time.

The crowd were shocked. They were astonished. They were flabbergasted. They had barely recovered their collective composure at witnessing *Rocket* hurtling along at speeds hitherto unimagined, and here was another machine travelling even faster! Coat tails flapping, the fearless engineers clung to their speeding machine, attending to its mechanical needs.

> It seemed, indeed to fly, presenting one of the most sublime spectacles of human ingenuity and human daring the world ever beheld. It actually made one giddy to look at it, and filled thousands with lively fears for the safety of the individuals who were on it . . .

Neither Braithwaite nor Ericsson had ever experienced anything like this before, and both were probably surprised at the remarkable performance of their engine. Not only was she fast, but her movement was as smooth as silk, in marked contrast to the rolling motion of *Rocket*:

> it passed the spectator[s] with a rapidity which can only be likened to a flash of lightning' [exclaimed *The Liverpool Chronicle*]. As it darted along under the grand stand, Messrs Braithwaite and Ericsson . . . took off their hats to the fair occupants, who acknowledged the compliment by cheers and waving of handkerchiefs . . .

Novelty captured imaginations along with hearts that day. She was the darling of the crowd, the people's choice, and few doubted she would capture the grand prize. Perhaps even George Stephenson's confidence was rattled by the remarkable performance.

Messrs Braithwaite and Ericsson did not run their machine for very long. There were still adjustments and last-minute things to take care of in preparation for her trials on the morrow, and they did not want to run the risk of a mishap. They cut off the steam supply and glided to a gentle stop. Their adoring crowd cheered on. They had just witnessed the unbelievable. Whatever would be next?

Some parts of Timothy Hackworth's *Sans Pareil* had been put together just that morning and she was far from ready to be demonstrated to the crowd. Hackworth, a rather dour-looking man of forty-three years, worked away on his machine in a state of much agitation. But the judges were very understanding. Earlier they had allowed him to make some test runs with the tender the company had supplied, just to check that the various parts of his machine were working properly. When he discovered that the boiler was leaking badly, a defect that would take several days to rectify, the judges granted him the extra time he needed to deal with the problem.

With *Sans Pareil* under repair, and Timothy Burstall's *Perseverance* temporarily out of commission, the last machine to be demonstrated to the crowd was Mr Brandreth's *Cycloped*. When the Rainhill trials were announced, Thomas Shaw Brandreth, a former director of the Liverpool & Manchester Railway Company, had written to his old board enquiring whether the competition was restricted to steam-powered locomotives. When the directors assured him they were open to any and all forms of motive power, he set to work designing and building his own machine.

Some of the spectators may have seen *Cycloped* before the demonstrations began. Others may have heard about it, and possibly learned it was not driven by steam. This may have raised speculation on its likely source of power. It might have been propelled by magnetism, or perhaps by the force of gravity. Perhaps it was some kind of perpetual motion machine. All manner of bizarre and incredible machines had been proposed in recent times for propelling vehicles, like Benjamin Cheverton's gas-powered engine that worked, seemingly without fuel, by alternately vaporising, then condensing, a sealed volume of gas. But none of these proposals had ever come into being – perhaps Brandreth's machine would prove the exception.

When *Cycloped* was unveiled before the expectant crowd it must have been an enormous anticlimax, especially after *Novelty*'s dazzling debut. But if the spectators were disappointed they appear to have hidden their feelings. And the newspaper coverage was anything but negative:

> Mr Brandreth's . . . engine [was] exhibited . . . as an exercise. About fifty persons clung round the wagons, giving a gross weight, with the machine, of about five tons . . . propelled . . . at the rate of five miles per hour. This could scarcely be called a fair trial of the ingenious inventor's machine, nor was it as such considered by the judges.

Significantly, *Cycloped* was not mentioned in the *Mechanics' Magazine*, save to say it was 'worked by a horse'. In fact it was worked by a pair of horses, moving on a treadmill geared to the driving wheels.

It is difficult to understand how a reporter who had seen locomotives travelling at unparalleled speeds could describe a horse-powered machine's inventor as 'ingenious'. Equally

Cycloped, *Brandreth's entry, was powered by a pair of horses, not a single one as suggested by this contemporary illustration.*

difficult to understand is why anyone should take the trouble to build such an anachronistic machine in the first place. But it must be remembered that this was an age when all land transport was powered by horses. Horses were an integral part of everyday life, whereas steam locomotives were a glimpse of the future. Horses certainly had primacy in Brandreth's life, so building a machine where their power could be used so effectively was probably quite ingenious.

Brandreth, a barrister by profession, had been actively involved in the Stockton & Darlington Railway, and was a staunch advocate of the use of horses on railways. He had taken credit for the 'dandy cart', a carriage that horses rode in on the downhill sections of the line to allow them to rest. This was not out of any concern for animal welfare but because horses rested in this way were able to haul greater loads during their working day. Although Brandreth claimed the dandy cart as his own invention, the idea had been suggested to him by George Stephenson. The horses used on that line could haul four heaped coal wagons, for a total load of twelve tons. The locomotives could haul between four and five times that amount, and at four times that speed. But only six locomotives were in use on the line at the time of Rainhill, compared with dozens of horse-drawn wagons. And horses were far more reliable than their temperamental counterparts. Horses therefore hauled a significant portion of the total tonnage of goods on the Stockton & Darlington Railway: 43 per cent in the year 1828. And all passengers were conveyed by horse-drawn vehicles. The continued use of horses on that line seemed assured, at least for the foreseeable future, and horses would endure on local lines until well into the following century. Regardless of the continued importance of horses to railways, *Cycloped*, with a top speed just half that stipulated in the contest rules, was irrelevant to the Liverpool & Manchester Railway and was not considered any further.

Cycloped was not the only vehicle exhibited at Rainhill that was muscle-powered. One Mr Winan exhibited a machine that was worked by two men. During its demonstration it conveyed six passengers, though at 'no great velocity'. Later on that afternoon it was involved in a collision with *Sans Pareil* in which one of its wheels was slightly damaged. Presumably there was no damage to Hackworth's sturdy machine.

As the sun began to dip in the west the first day of the Rainhill trials drew to a close. Not that any trials had taken place, but the spectators had been treated to a momentous and thoroughly thrilling day. The unquestionable highlight of the day had been the spectacular performance of *Novelty*. And as the spectators drifted from the grounds for their various homes or accommodation, they could look forward to the morning and seeing their favourite undergoing her trial.

Although the day had been rather uneventful from the judges' perspective – there had been no disputes to settle and they did not have to make any major decisions – it did give them the opportunity of considering how to assess the performance of the competitors. On their return to Liverpool that evening the three men met together to work out the details of how this would be done. And 'after the most mature deliberations we decided upon the following Plan . . .'. The procedures they devised were to help them judge how the locomotives would perform on a daily basis in travelling between Liverpool and Manchester.

At eight o'clock on the morning of its trial the locomotive would be weighed, with its boiler full of cold of water. Its assigned load would be three times that weight. The tender would then be attached and loaded with as much fuel and water as the operator thought sufficient for a journey of thirty-five miles. The entire weight of the tender and its contents would be considered as part of the assigned load. Engines that carried their own fuel and water, namely *Novelty* and Burstall's *Perseverance*, would be allowed a proportionate deduction from their assigned load, according to their weight.

The locomotive, coupled to its assigned load, would be pushed up to the starting post by hand. The amount of fuel required to raise steam would be weighed into the empty fireplace, and lit. The time taken for the steam pressure to reach 50 psi (pounds per square inch) would be noted, and the trial would begin.

The distance between the starting and finishing point of each trip was set at one and three-quarter miles. This included one-eighth of a mile at either end for speeding up and slowing down, so the locomotive could travel one and a half miles at full speed. Ten return trips would be made, for a total distance of thirty-five miles, which was the distance between Liverpool & Manchester, thirty miles of which would be a full speed. The locomotive would then take on more fuel and water for a second set of ten trips. The times for each trip would be accurately recorded, as well as the time taken to prepare for the second thirty-five-mile journey. If a locomotive ran out of fuel or water during its journey the time taken to replenish these would be added to its travel time. The regulations were to be printed on cards for circulation to the competitors and directors.

Messrs Rastrick, Wood and Kennedy could retire to their beds feeling well satisfied. Having produced a set of rules for the contest was a major accomplishment. But these were not the only set of regulations governing the competition. From the outset, the directors of the Liverpool & Manchester Railway Company had laid down a set of stipulations with which all contestants must comply, and the judges had been careful to ensure their rules did not interfere with these in any way. These stipulations, which were quite detailed, were based on lessons learned during the operation of the Stockton & Darlington Railway. Since some board members served on both railways, there was a free exchange of ideas and information between the two companies. Some of the stipulations, like the weight restriction, were to conserve the track, and the wheels. Others, like the requirements for safety valves, were to protect lives and property. But when it came to safety issues, the locomotive drivers themselves were a greater liability than the engines they drove.

CHAPTER 2

Lessons from the past

The early locomotive drivers were an elite group of men. They were a rare breed too because locomotives were so uncommon. And their job of handling the difficult and sometimes lethal machines in their charge set them apart from other manual workers. It was said that 'only one man in three could make a driver, because the engines were so difficult to manage . . .'. The drivers were very well rewarded for their skills, and they were among the highest paid of the working classes.

In the early days of the Stockton & Darlington Railway, drivers were paid by how much freight they could haul, so most of them would cut any corner to save time. George Sunter could run his train of wagons from Shildon to Middlesbrough without stopping, even though he had to make two water stops along the way. That section of the line was downhill, and he accomplished this feat by uncoupling the wagons and dashing off with the locomotive at full throttle, allowing them to coast down the incline. By the time he had taken on water the wagons had caught up, and he could be on his way again. The company were highly disapproving of such cavalier behaviour and Sunter would doubtless have lost his job had he been caught in the act. But other seemingly dangerous practices were tolerated. The wagons, for example, had to be lubricated while the train was in motion:

> This was done by having an oil tin and brush . . . and was always done when the line was pretty level. The driver regulated the speed before getting off to about two miles per hour, but often before the whole train was oiled the speed got up to four miles per hour on account of being lubricated and the driver and fireman had to run the whole length of the train to get to the engine. In the case of a loaded train they got to the last wagon and walked over the tops of the other wagons to the engine.

The earliest railway tracks were made of cast iron, which is quite brittle, and the weight of the locomotives frequently caused them to break. George Stephenson therefore recommended the Stockton & Darlington Railway should use wrought iron, which is much tougher. But the company continued using cast-iron wheels, and these broke with alarming frequency, largely because of the unevenness of the track. This was especially so during the extreme cold, when upward of 1,500 wheels would break every month, presumably because the frost caused the ground to heave. Wheel breakages were exacerbated by the complete absence of springs, which is why springs were stipulated for the Rainhill contestants, alongside the weight restriction. Increased speed was also a contributing factor, and the company accordingly imposed a strict speed limit of 8 mph. Drivers caught exceeding this limit faced dismissal.

But broken wheels were naught compared with the havoc a driver could wreak by allowing a boiler to explode.

On 19 March 1828 John Gillespie, a novice fireman, was taking his first journey aboard *Diligence*, driven by James Stephenson, elder brother of George Stephenson. It was the fireman's first day on the job, and he was under the watchful eye of the retiring fireman, Edward Corner, who was making his last journey. They stopped somewhere between Shildon and Darlington to pump fresh water into the tender. There was a strong wind blowing in the direction of the furnace, and this caused the fire to burn more brightly.

The early locomotives were often poor at maintaining steam pressure when they were in motion. However, when stationary and no longer using steam, the pressure built up, and the function of the safety valve was to vent the excess steam when it reached a certain pressure, usually about 50 psi. But some drivers routinely held the safety valve down when the engine was stationary, to build up an extra head of steam, though they were expressly forbidden from doing so by the company. Such was the case on this occasion, and while there is no official record as to who was responsible, it must have been Stephenson because he was the driver. This would also accord with his unruly reputation, which often got him into trouble for some infringement of company rules.

Holding down Diligence's safety valve would have been dangerous enough under normal circumstances, but with a strong wind fanning the furnace it was lethal. The steam pressure rapidly built up until the boiler was no longer able to withstand the mounting stresses, and there was an enormous explosion. The blast threw the retiring fireman sixteen yards, breaking his thigh. John Gillespie's wounds were far more serious, but Stephenson escaped uninjured. After the accident Thomas Brandreth presented the old fireman with half a sovereign (£29) in compensation. John Gillespie received one whole sovereign, but he died from his wounds before he had a chance to spend it. There was no inquiry, and no blame was attached to Stephenson for the accident. Nor were any lessons learned, either by the company or by the drivers, because, just three months later, *Locomotion* exploded under similar circumstances, killing the driver, John Cree. Significantly, perhaps, these catastrophes hardly got any mention in the company records. Hackworth, who was in charge of all locomotives, merely recorded:

> July 1 – 1828 John Cree goeing [*sic*] down the line No 1 Locomotive at Aycliffe Lane while getting water the Engine exploded . . . Cree died on the 3 at 3 o-clock morning.

The next item is a routine entry about the delivery of a piece of equipment, and the accident is not mentioned again.

Locomotive explosions were by no means confined to the Stockton & Darlington Railway. Ten years before the disasters of *Diligence* and *Locomotion*, a colliery engine named *Salamanca*, built by locomotive pioneer John Blenkinsop, underwent a spectacular explosion. Her driver was waiting for the delivery of some empty wagons, and, despite previous warnings of the perils of the practice, he allowed the steam pressure to rise dangerously high. Like the operators of the two previous locomotives, he had tied the safety valve down. Unable to withstand the pressure, the boiler exploded, throwing the driver into an adjoining field. When they found his body, one hundred yards away, 'it was in a very mangled state . . . and it was quite dead.'

Ruptures of the boiler barrel itself were exceedingly rare, and almost all explosions involved the catastrophic failure of the flue tube, invariably at the furnace end, where it was riveted to the end of the boiler. As mentioned earlier, straight-through flues ran through the boiler barrel and were continuous with the chimney at the front, and with the furnace at the back. The furnace was merely the rear end of the flue tube – two or three feet in extent – fitted with iron bars to form a grate for the fire, and closed by a hinged fire door. When the fire was lit and the flue tube got hot it expanded, but was held in place at either end by the boiler end-plates. This placed large stresses on the union between flue tube and end-plate, especially at the much hotter furnace end. This was exacerbated by the uneven heating of the flue, the top becoming much hotter than the bottom, causing banana-like flexing. Uneven heating was an inherent problem of flue-tube boilers, a consequence of the poor heat circulation in the water. Before the working replica of *Locomotion* was modified, heat circulation was so poor that the top of the boiler could be hot to the touch while the bottom was still stone-cold.

The flue tube was open to the air at both ends – via the furnace at one end and the chimney at the other – and so was at atmospheric pressure. But the surrounding water in the boiler was at the pressure of the steam pressing down upon it. Consequently, when the flue tube failed, there was an explosive burst of high pressure steam and water from the boiler barrel into the furnace. Death and injury were caused by searing steam, scalding water and red-hot coals. The fireman was in the direct line of fire, but the driver, perched high on his platform beside the boiler, usually escaped serious injury. As the explosion was confined to the flue tube and end-plate, the locomotive itself was usually repairable. In an attempt to eliminate boiler explosions, the directors of the Liverpool & Manchester Railway stipulated that the locomotives entered for the Rainhill trials must have two safety valves, one of which had to be out of the driver's reach, to prevent tampering.

Although steam had been used for generating power since the early eighteenth century, boiler explosions were quite rare until the early part of the nineteenth century. This was not because machine operators had become any less responsible, but because of a radical advance in technology. This breakthrough was brought about by the use of high-pressure steam, and, without such a change, steam locomotion could never have come into being.

The earliest engines operated at steam pressures of only a few pounds per square inch. They were great ponderous behemoths, standing taller than most houses. Each had a horizontal wooden beam, as large as a footbridge, which was rocked up and down by a gigantic piston. The single cylinder, typically four or five feet in diameter, could have held several people and was probably roomier than a slum tenement. Although these beam engines are usually referred to as steam engines, they are more correctly called atmospheric engines, because they operated on an entirely different principle. In steam engines proper it is the pressure of the steam acting upon the piston that produces the force, and this drives the piston towards the other end of the cylinder. In atmospheric engines – also known as 'fire engines' – the cylinder was filled with steam at such low pressures that it produced little, if any, force upon the piston. But when the steam cooled it condensed into water droplets, and since these occupied a tiny fraction of the volume formerly

The frock-coated gentleman in this 1717 drawing of a Newcomen engine gives an idea of its size.

occupied by the steam, a partial vacuum was created. The pressure inside the cylinder was now much less than that outside, so the piston was pushed down by atmospheric pressure.

The principle of the atmospheric engine can be demonstrated by a simple but rather impressive experiment with a tin can. I have a vivid schooldays memory of our science teacher performing the experiment before a class of wide-eyed twelve-year-olds. Taking a flat-sided gallon can with a screw-top lid, he added a little water and placed it over a Bunsen burner to boil. When it was steaming away vigorously he replaced the lid, screwed it down tightly, and put the can down on the lab bench. We all waited expectantly, but nothing happened at first. Then, to our collective astonishment, there was an impressively loud crunch as the can was crushed almost flat. The experiment can be repeated in the kitchen using a wide-mouthed plastic juice container and a kettle. With the mouth inverted over the spout of a steaming kettle the container is filled with steam until it billows out freely. Once the lid has been screwed on tight it takes only a few seconds before the atmospheric pressure begins crushing the container. This can be made to happen almost instantly by dousing the container with cold water.

In the engines of the early 1700s, designed and developed by Thomas Newcomen (1664–1729), the steam was condensed by spraying a jet of cold water into the cylinder. The cylinder was open to the atmosphere at the top, and the connecting rod of the piston was attached to an overhead beam by a chain. The other end of the beam was attached, by another chain, to a weighty rod, made sufficiently heavy to raise the piston. As the piston ascended, steam was admitted to the closed bottom end of the cylinder, by a valve. Once the piston reached the top of the cylinder the valve was closed, and the water jet was sprayed inside, rapidly condensing the steam. The atmospheric pressure above the piston was now greater than the partial vacuum below, forcing the piston down. Just before the piston reached the bottom of the cylinder at the end of its power stroke, a valve opened at the bottom, to eject the condensed water. As the piston began its ascent under the counterweight of the rod, the steam valve opened again, and another cycle began. The beam rocked back and forth like a seesaw, and the reciprocating motion was used to drive a water pump, steam power having originated specifically for pumping water from flooded mines.

The first Newcomen engine was built in 1712, for a colliery in Staffordshire. Most of Newcomen's other engines were erected in the coal-mining districts of the Midlands and the North, where their prodigious appetites for fuel could be satiated without direct costs to the operators. In addition to paying for the cost of the engine, the mine owners had to pay royalty fees for its use, and this additional burden probably slowed the spread of the new invention. By the time the patent expired in 1728 there were fewer than one hundred Newcomen engines, but the number grew to almost 1,500 by the end of the succeeding century. The earliest machines had modest-sized cylinders, seldom exceeding 2 ft in diameter, and were usually cast in brass. But as the demand grew for more powerful engines, to pump deeper and bigger mines, size increased. Brass, which was much costlier than iron, became prohibitively expensive, and was soon replaced by iron castings. Initially there were no mechanical means of smoothing cylinder bores, so it had to be done by hand. The inside of the cylinder was coated with a mixture of oil and emery powder, and a team of men would then drag a solid lead plug back and forth on a rope, rotating the cylinder fractionally as the bore was worn smooth.

By the 1760s atmospheric engines were being built with cylinder diameters in excess of 6 ft. Such gigantic engines could be used to pump water from the deepest mines, but they were enormously expensive on fuel because their efficiency was so low (about 0.5 per cent). The primary reason was the enormous heat loss when cold condensing water was sprayed inside the hot cylinder prior to each power stroke. James Watt (1736–1819) realised this after experimenting with a working model of a Newcomen engine. At that time (1763–64) he was working as a laboratory technician at the University of Glasgow, and had been given the model to repair for a class. A year or two earlier he had experimented with high-pressure steam by connecting a small syringe to an apparatus like a pressure cooker. Surprised at the considerable force exerted on the plunger, he thought it might be an effective way to utilise steam. 'But I soon relinquished the idea of constructing an engine upon its principle . . .' he wrote, primarily because of 'the danger of bursting the boiler.' Watt remained resolutely opposed to using high-pressure steam for the rest of his life.

Convinced that the only safe way of generating power was to use low-pressure steam, Watt pondered over how to improve Newcomen's engine. Two years later he arrived at an elegant solution. If a small and separate cylinder were connected to the main cylinder by a pipe, and if cold water were sprayed inside it, the steam would be condensed. And, since the steam in the main cylinder was 'an elastic fluid, [it] would immediately rush into the empty vessel [the small cylinder], and continue so to do . . . until the whole was condensed'. Spraying cold water into a separate condenser should therefore have the same effect as spraying water into the main cylinder, but without the disadvantage of cooling it down. The working model he built confirmed his conviction.

Watt formed a partnership with an astute businessman and factory owner named Matthew Boulton (1728–1809), who was also a talented engineer. This was only after Watt had taken out a patent for the improved engine with a financial backer, in 1769. When the backer subsequently went bankrupt Boulton bought his shares in the patent, clearing the way ahead. Watt had yet to build a full-size working version of the engine, and was still struggling towards that end six years after obtaining the patent. His biggest problem was getting a good seal between the piston and the cylinder, engineering tolerances being so poor at the time. Then Boulton learned that John Wilkinson (1728–1808), an eminent ironfounder, had recently patented a machine for boring cylinders with unparalleled accuracy – largely copied from an existing design. The cylinder Watt had been working with was scrapped and Wilkinson was commissioned to build another. Watt assembled the engine in Boulton's factory. The new cylinder was a considerable improvement, and to get a good seal for the piston Watt used rings of hemp and oakum (teased rope), lubricated with tallow.

The engine was a great success and orders flooded in. The largest market was in Cornwall, for pumping water from flooded copper and tin mines. Being so far from the collieries coal prices there were at a premium, and purchasers of the new engine were charged a levy, based upon their fuel saving over the wasteful Newcomen engines. But mine owners became increasingly resentful, and some enterprising mine captains – as the managers and engineers were called – began erecting their own engines to avoid the royalties. Boulton & Watt prosecuted violators to the full extent of the law. One of those with whom they had legal dealings was Richard Trevithick, the man who initiated the great leap forward from low- to high-pressure steam engines.

Trevithick was a powerfully built Cornishman with clear blue eyes and a free and lively spirit. He was 6 ft 2 in. tall, with broad shoulders, and it was said he could throw a 14 lb hammer clean over the engine-house chimney of the tin mine where he worked as a youth. He had a natural talent for machines, and was soon erecting atmospheric engines in mines throughout much of Cornwall. At the age of only twenty-two he was called as an expert witness on behalf of William Bull, a fellow Cornishman who was being sued by Messrs Boulton & Watt for patent infringement. Four years later he was facing legal action himself, for a patent infringement involving an engine he erected at a mine near Penzance. But he does not appear to have been unduly worried by it all, even though Boulton & Watt had successfully sued several other Cornishmen: he was rather preoccupied with an invention of his own.

The idea of using high-pressure steam to drive a piston had apparently occurred to Trevithick quite spontaneously, some time before 1797. The singular advantage of high-pressure steam is that far greater forces are generated for a given piston area, so engines can be considerably smaller. Trevithick was a hands-on practical man with no formal academic training, and he wasted little time in developing his idea. By about 1797, which was also the year of his marriage, he had the first of several working models. In later years his wife, whose family, the Harveys of Hayle, owned a foundry that built steam engines, often spoke of how these models were tried out in their Camborne home. One evening he invited their neighbours, Lord and Lady de Dunstanville,

Richard Trevithick.

who owned several local mines, to witness the first model in action. The gleaming brass engine was built to Trevithick's specifications by his wife's brother-in-law, an accomplished metalworker associated with the Harvey foundry. It had a single cylinder, connected to a flywheel to continue the motion of the piston beyond its fully extended position; having no flywheel in a single-cylinder engine would be like trying to pedal a bicycle with only one foot.

A second model soon followed. This one had an integrated boiler and was mounted on three wheels, one of which was linked to the piston. As in the first model, there was a single piston, linked to a flywheel. Steam was delivered to the cylinder by a valve which alternately supplied either end, thereby making each stroke a power stroke. Anybody seeing the diminutive engine careering around the floor of the Trevithicks' Camborne home would have thought it unique. But some similar experiments had been carried out in the neighbouring town of Redruth when Trevithick was a boy.

Redruth, like Camborne, was at the centre of the Cornish copper and tin mining industry and William Murdock was employed there by Messrs Boulton & Watt to superintend the erection of engines. These engines, of course, all worked at low pressures, but Murdock experimented with high-pressure steam during his spare time. The very idea of using high-pressure steam, with the attendant risk of boiler explosions, was anathema to Watt, so Murdock had to conduct his experiments clandestinely. Aside from being outstandingly creative – Murdock's inventions include using coal gas for lighting – he was skilled with his hands, and built a model steam carriage, probably in 1784. This was much like the one that would run around Trevithick's home more than a decade later. In spite of Murdock's attempts at

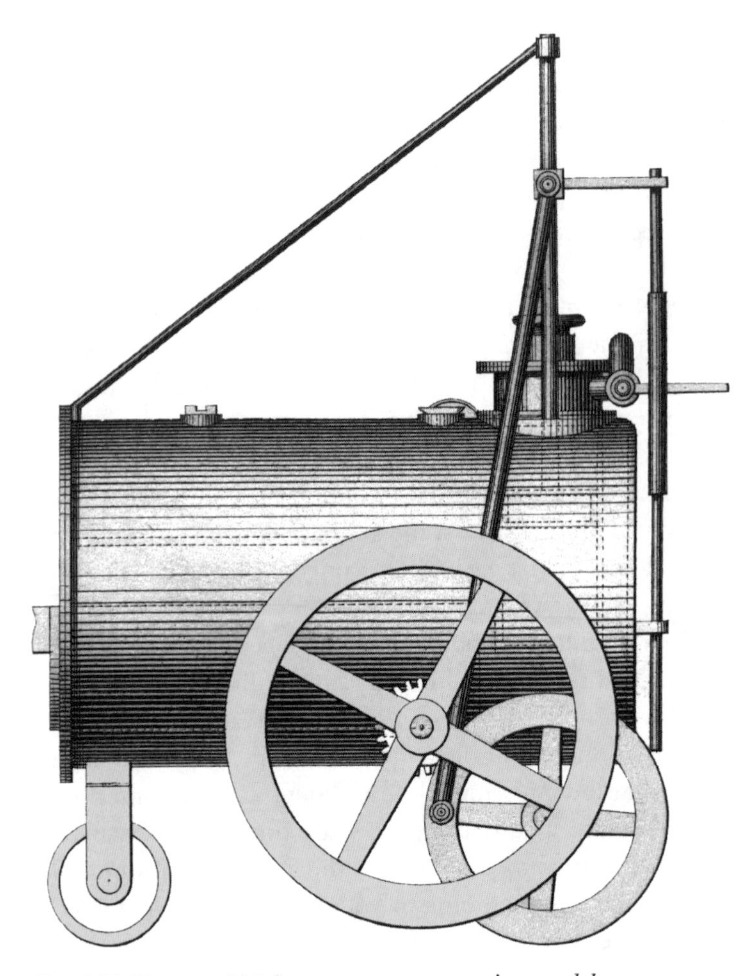

Trevithick's second high-pressure steam engine model.

secrecy his employers found out about his experiments. They were appalled, and persuaded him to desist.

There is some question regarding the extent to which Trevithick was influenced by Murdock's earlier experiments with high-pressure steam. The two certainly knew each other, and Murdock's son wrote a letter stating that his father had shown the working model to Trevithick, about two years after its construction. Trevithick was then only fifteen, and it is arguable how much the experience influenced him in later years.

By about 1799 Trevithick was selling portable high-pressure steam engines for use in mines. Rather than going into production himself he contracted out the manufacture of his engines to various workshops. These were built in different sizes, according to power requirements. They operated at a pressure of 25 psi, and some had twin cylinders. The cylindrical boiler had a return flue, with a tall chimney. The single cylinder was set into the top of the boiler to keep it hot. The piston rod projected vertically from the top of the cylinder and pushed against a horizontal cross-head. This moved up and down with the movements of the piston, guided by a pair of vertical rods. A connecting rod linked the cross-head with a large flywheel, and the whole engine was firmly anchored to a base. Unlike the massive atmospheric engine that required a brick wall to support its heavy beam, the high-pressure engine was self-contained and compact. The smaller ones were quite portable and could be readily loaded on to the back of a wagon. An engine of this type, built around 1805, can be seen on display in London's Science Museum. This collection also houses a model steam carriage, probably built a year or so earlier, which closely matches the descriptions of Trevithick's second model.

The high-pressure engines were known as 'puffers' for the distinctive sound of their exhaust steam, which apparently carried for miles across the rugged Cornish countryside. They were first used to replace the hand windlasses and the horse-powered winding engines, or whims, used for hauling minerals to the surface from the depths. There was a ready market for the new engines, and Trevithick's account book showed many orders. Significantly, his accounts never balanced, underscoring his ineptitude, or lack of interest, in business matters.

By the turn of the eighteenth century Trevithick was supplying equipment to many of the mines in the area. They had evocative names, like Wheal Druid, Ding Dong and Cook's Kitchen, and large sums of money were passing through his hands. In all, he sold about thirty high-pressure engines, which is not a huge number, but the Cornish copper mines were going through a depression at the time. If Trevithick had been properly focused his high-pressure steam engines would likely have made him a wealthy man. But his agile mind was far more occupied with new projects than with the more mundane business of filling orders and finding new customers. What Trevithick badly needed was a man like Watt's Boulton, somebody to take care of business while he got on with invention.

At the time his portable steam engines were replacing horse-driven whims Trevithick was exploring the possibility of building a full-size steam carriage. He began by conducting a simple experiment to see whether a carriage could be propelled by turning its wheels. He and his friend and confidant, Davies Gilbert, parked a carriage on a hill, unhitched the horse, then pushed down on the spokes of the wheels with all their might. They were pleased to discover that the carriage moved forward, up the hill. Viewed in hindsight, such an outcome is a foregone conclusion, but we have to slip our third-millennium feet into their Georgian shoes to grasp the significance of the experiment. Back then all wheeled vehicles were pulled, or pushed, the wheels rotating passively: in no case was the motive force applied directly to the wheels. It was therefore by no means certain that there would be enough traction between the rim and the ground to prevent the wheel from slipping, leaving the carriage where it stood. Demonstrating that the application of force to a wheel produced propulsion cleared the way ahead. The traction experiment was probably carried out in 1800, which was the year Trevithick began building the Camborne steam carriage.

The carriage was assembled in the workshop of the local blacksmith, where many of the smaller parts were made. The castings, and the plates for the boiler, were supplied by the Harvey foundry. Like the blacksmith's workshop it was poorly equipped: 'there were but a few small hand-lathes fixed on wooden benches, a few drilling machines, and but one chuck lathe'. The lack of precision in the manufacture of its many parts made for difficulties in the final assembly, completed some time towards the end of 1801. Although there are several eye-witness accounts of the first test runs, there are no good descriptions of the carriage itself, and the illustrations given in Trevithick's biography, written by his son, are quite unreliable. In all probability it looked like a converted horse-drawn carriage, rather than a wheeled version of a portable steam engine, as depicted by his son. The first trial took place on Christmas Eve 1801, as remembered by Stephen Williams:

> Captain Dick got up steam, out in the high-road just outside the [blacksmith's] shop at the Weith. When we see'd

> that Captain Dick was agoing to turn on steam, we jumped up as many as could; may be seven or eight of us. 'Twas a stiffish hill going from Weith up to Camborne Beacon, but she went off like a little bird.
>
> When she had gone about a quarter of a mile, there was a roughish piece of road covered with loose stones; she didn't go quite so fast, and as it was a flood of rain, and we were very squeezed together, I jumped off. She was going faster than I could walk . . .

While this was the first steam carriage witnessed in England it was not without precedent, because a Frenchman, Nicholas Cugnot, had built a steam wagon in 1769. Cugnot was an artillery officer and his machine was primarily for military use. It worked fairly well, if only at a walking pace, but its boiler appears to have been less adequate at maintaining steam than Trevithick's, and it was soon abandoned. However, whereas Cugnot's machine was preserved for posterity, the Camborne engine suffered a funereal fate, just four days after its inaugural run.

After 'travelling very well' it broke down, and was pushed under a shelter while Trevithick and his companions 'adjourned to the hotel, and comforted their hearts with a roast goose, [and] proper drinks . . .'. Unfortunately they had forgotten to extinguish the fire, and when the boiler ran dry 'the iron became red hot, and nothing that was combustible remained, either of the engine or the house'.[21]

Trevithick seems to have been undeterred by the setback. Over Christmas dinner he and his cousin, Andrew Vivian, became partners, and made plans to visit London to patent the steam carriage, and the high-pressure steam engine. Their timing was propitious because Boulton &

Watt's patent had expired the previous year, leaving the field wide open for others.

Their sojourn in London was an unqualified success and they obtained their patent within two months of applying. Acting on advice received during the trip, they began work on a second road carriage, to be demonstrated in London. Meanwhile Trevithick visited Coalbrookdale, in the Midlands, at the very heart of Britain's iron industry, to work on a new high-pressure engine. This one ran at the staggeringly high-pressure of 145 psi – the highest pressure that any steam engine had ever operated – but he intended going even higher: 'The boiler is 1½ inch thick, and I think there will be no danger in putting it still higher. I shall not stop loading the engine until the packing [around the piston] burns or blows out under its pressure.'

The cylinder had a diameter of 7 in., comparable to his portable engines, but small compared with beam engines which were typically 4 ft in diameter. These giants were capable of generating forces of six tons or more and the Coalbrookdale engineers were therefore amazed when his tiny engine did about half as well. 'If I had fifty engines,' Trevithick wrote to his friend Gilbert, 'I could sell them all here in a day, at any price I would ask for them. They are so highly pleased with it that no other engine will pass with them.'

Regardless of his optimism over this new high-pressure engine, nothing seems to have come of it, and there is no evidence that he sold a single one. Perhaps it was simply too far ahead of its time to be developed with the limited engineering capabilities then available. In any event, he spent little or no more time on the project before moving on to the next, which was typical of the man. Aside from continuing with his regular high-pressure engines, and working on the second steam carriage, his active mind was already turning to steam locomotion. A steam locomotive was reputedly built at Coalbrookdale, to his specifications, sometime between 1802 and 1803, but there is no convincing evidence.

The London carriage was demonstrated in the capital in the spring of 1803, to a mixed reception. Many residents cheered and waved at the unprecedented sight of a steam engine puffing and clattering along the streets. But some omnibus and cab drivers, concerned for their livelihoods, pelted them with cabbage stumps and eggs. But their fears were unfounded because the roads, like those elsewhere, were so atrocious that a steam carriage could not go far without being damaged or upset. Trevithick was among the first to realise that if steam locomotion had any place in Britain in the early days of the nineteenth century, it was only on a railway track.

The London carriage, Trevithick's second steam engine for running on roads.

Their steam carriage 'ran well . . .' but, as Vivian recounted, 'there was some defect, and our finances being low, we were compelled to abandon it'. Regardless, Trevithick's future appeared bright. Shortly before the demonstration he had installed an engine for the Admiralty for boring cannon. Britain, embroiled in the Napoleonic Wars since 1793, had a great need for ordnance, and the Admiralty was so pleased with the engine it planned to purchase several more, declaring that no other would be used in the government's service. Trevithick had also sold a quarter share of his patent, to an unnamed gentleman, for the huge sum of £10,000 (£450,000). Writing to Gilbert about developments, he further announced that 'I have a prospect before me [in south Wales] of doing exceedingly well.'

The only dark cloud on Trevithick's blue horizon was occasioned by the devastating explosion of one of his stationary engines. The inexperienced operator left in charge of the engine had brought it to a stop without releasing the safety valve, which had been held down. The inch-thick boiler exploded in a deadly shrapnel of cast iron, bricks and mortar, instantly killing three people and mortally wounding a fourth. One boiler fragment, weighing about 250 lb, was thrown more than 300 ft. Such an enormous explosion received a good deal of attention, but a great deal more was made of the situation because it was one of Trevithick's high-pressure engines. His old adversaries, Messrs Boulton & Watt, made great capital of the disaster, doing 'their utmost to report the explosion, both in the newspapers and in private letters . . .'.

This was Trevithick's first boiler failure and he was determined it would be his last: 'I shall put two steam [safety] valves and a steam-gauge in future, so that the quicksilver shall blow out in case the valve should stick, and all the steam be discharged . . .'. He never experienced another explosion.

The unnamed gentleman to whom Trevithick had sold part of his patent share was Samuel Homfray (1761–1822), who owned an ironworks at Pen-y-darren, just outside Merthyr Tydfil in South Wales. He used a number of steam engines, all of the Boulton & Watt pattern, and was therefore very interested in Trevithick's high-pressure engines. The two men got along well together, and Trevithick wrote to Gilbert that 'Mr Homfray . . . has taken me by the hand . . .'. Homfray was wealthy and influential, and succeeded in blocking an attempt by Boulton & Watt to have a Bill introduced in Parliament banishing high-pressure steam engines on the grounds of public safety.

An iron wagonway for horse-drawn wagons linked the Pen-y-darren Iron Works to the Glamorganshire Canal, some nine miles away, and Trevithick easily persuaded Homfray to have him build a steam locomotive to replace the horses. Working with local craftsmen, Trevithick completed the locomotive early in 1804. Discounting the Trevithick-designed locomotive which may have been built at Coalbrookdale the year before, this was the first locomotive in the world. Homfray was a betting man, and made a 500 guinea (about £24,000) wager with a neighbour that the locomotive would haul a ten-ton load of iron the entire length of the wagonway.

The engine was first tested on the track on Monday 13 February 1804: 'It worked very well, and ran up and down hill with great ease, and was very manageable. We had plenty of steam and power . . . The bet will not be determined until the middle of next week . . .'.

Looking like something from a clockmaker's nightmare, the Pen-y-darren engine was a whirring complex of

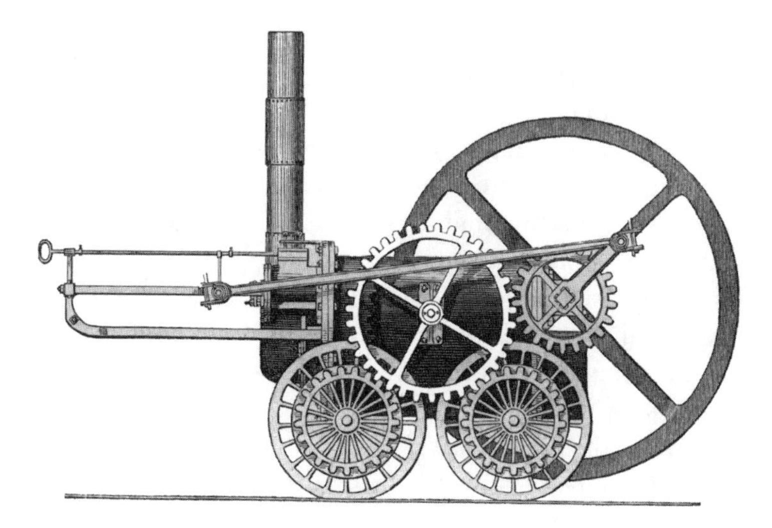

Trevithick's Pen-y-darren locomotive.

cogwheels and sprockets, with an enormous flywheel, 8 ft in diameter, driven by a single horizontal piston. Its 8 in.-diameter cylinder was set deep into the return-flue boiler, and an extensive cagework extended forward from the boiler, forming a guide for the cross-head attached to the piston rod. The cross-head was coupled to a pair of long connecting rods that turned the flywheel on its axle. The piston stroke – the maximum distance the piston moves – had to be correspondingly long, and was 4 ft 6 in. A cogwheel attached to the flywheel's axle meshed with a larger cog-wheel, centred between the two pairs of driving wheels. This meshed with the cogs on the driving wheels, turning them as the flywheel turned. Because of the reduced gearing, the smooth-rimmed driving wheels rotated more slowly than the flywheel. There was no footplate, and the operator probably walked alongside the machine as it trundled along the tramway. With the 6 ft-long boiler filled with water the engine weighed five tons.

As Trevithick explained to Gilbert, the 'steam that is discharged from the engine is turned up the chimney . . . The fire burns much better when the steam goes up the chimney than when the engine is idle . . .'. This clearly shows that Trevithick had discovered that the intermittent blasts of exhaust steam from the cylinder, when discharged into the chimney, made the fire burn more fiercely. This was because the steam accelerated the flow of exhaust gases, decreasing the pressure inside the chimney, thereby improving the draught. Its effectiveness would be increased in later locomotives by narrowing the diameter of the discharge pipe, forming a device called the blast pipe. The blast pipe was eventually used on all locomotives, and a great controversy ensued over who was its inventor. Hackworth and Stephenson were the leading contenders, but the honour clearly belongs to Trevithick.

The engine easily hauled its ten-ton load, winning the wager, though Homfray's neighbour reneged on payment. Trevithick wrote to Gilbert that the 'public until now called me a scheming fellow, but now their tone is much altered.' *The Cambrian* newspaper concluded its reporting of the event with the prediction 'that the number of horses in the kingdom will [now] be very considerably reduced . . .'.

This prediction would not come true for several more decades, and the Pen-y-darren engine had a short and undistinguished career. The failing had less to do with the locomotive than with the wagonway, which frequently broke under its great weight. On one particular journey 'she broke a great many of the tram-plates' and finished up by running off the track. She 'was brought back to Penydarran [*sic*] by horses [and] was never used as a locomotive after this . . .'.

A second locomotive, almost identical to the first, was built for Christopher Blackett, owner of the Wylam Colliery, near Newcastle upon Tyne. Trevithick appears to have supplied the drawings, but that was his only involvement. John Steel, who had worked with Trevithick, undertook much of the construction, which took place at Gateshead, near Newcastle. The work was completed, probably in 1805, but Blackett refused to accept delivery, probably because he thought the locomotive would destroy the wooden track. Like its predecessor, the locomotive was relegated to use as a stationary engine to drive machinery.

Towards the end of that year (1804) Vivian's account books showed they had received £1,250 (£67,000) in patent premiums, at an expenditure of £1,097 (£59,000). Part of the reason for the small profit margin was attributable to defaulters, which is ironic given Trevithick's previous cavalier attitude towards patents. The poor returns may have been the reason why his association with Homfray came to an end. But Trevithick still had his £10,000 (£450,000) – or what was left of it after financing his other business enterprises.

The boldest of these was a new locomotive, named *Catch-me-who-can*, which he exhibited in the heart of London. According to an early announcement in *The Times*, the engine was 'preparing to run against any mare, horse or gelding . . .'. The odds of 10,000 to one were in favour of the locomotive, whose 'greatest speed will be 20 miles in one hour . . .'.

Trevithick charged 5*s* (about £12) admission, and a high perimeter fence around the circular track concealed the spectacle from non-paying eyes. Judging from the press coverage the irrepressible showman spared no effort to promote the event, but although it attracted some attention

Catch-me-who-can, *exhibited in London in 1808, intrigued and amused, but the crowds did not flood to the fenced arena in the numbers Trevithick had expected.*

the crowds did not flood in as he had hoped, and certainly not in sufficient numbers to offset the costs. Nor does any serious attention appear to have been paid to the exhibition. The locomotive was little more than an intriguing curiosity, to be viewed with the same level of interest as a crocodile from Africa or a Red Indian from America. Nobody in London, it seemed, recognised the economic potential of steam locomotion. But the industrial north, with its immediate needs to transport minerals from the mines, saw the locomotive as a promising alternative to horse haulage, which was becoming increasingly expensive.

After only a few weeks in operation a rail broke, derailing the locomotive. This was the last straw for Trevithick and he promptly closed the exhibition. His experience in London was just like that in Wales. Recalling the Pen-y-darren trials eight years after the event (1812), Trevithick wrote:

> I thought this experiment would show to the public quite enough to recommend its general use; but though promising to be of so much consequence [the locomotive] has so far remained buried, which discourages me from again trying at my own expense, for the public . . .

Catch-me-who-can was Trevithick's last locomotive. When Blackett wrote the following year asking him to build a locomotive for his Wylam railway – now relaid with iron track – Trevithick declined, saying he was engaged in other pursuits. The experience of the exhibition in London had left him completely demoralised. He had even disposed of some of his equipment when the exhibition closed, because a working model of his high-pressure engine appeared in a shop in Fitzroy Square, less than a mile from the *Catch-me-who-can* circuit. And there it remained for some long time. Then, quite by chance, it was purchased by a foreign gentleman, setting in motion a chain of events that would change Trevithick's life for ever. Since these remarkable events have a bearing on the Rainhill story, their retelling is worth the while.

In the spring of 1813 a sailing vessel out of Jamaica was making for the Cornish port of Falmouth. She had been at sea for several weeks, and the passengers and crew were looking forward to sighting land again. One passenger in particular was anxious to set foot on *terra firma*. This was the last leg of a six-month passage from Peru, a journey that had not agreed with his health, which was now in a parlous state. But he had been ill before, and would doubtless be ill again, such were the perils of living in the tropics. Don Francisco Uville would not let his illness stand in his way – he was a man on a mission. He made no secret of his quest, and from the outset had willingly shared his story with any passenger interested enough to listen.

Uville explained he was an agent, representing certain mining interests in Lima. His listeners may have drawn a little closer when he began regaling them with stories of the treasures of the Peruvian Andes – gold, silver, platinum, copper, mercury, lead – more riches than any man could dream of. But so many of the mines had fallen into disuse because of flooding. Having heard of the remarkable achievements of English engineers in draining flooded mines, he had journeyed to England two years earlier, hoping to purchase the necessary steam engines and pumps. Money was no object, but he had been singularly unsuccessful in his quest. The great Boulton & Watt had gloomily dwelled upon the insurmountable difficulties of transporting heavy machinery across narrow mountain tracks. Besides, their atmospheric engines would be much less effective in the high Andes, where the atmospheric pressure was half that in England which lies mostly close to sea level. They could offer neither help nor hope. Wandering disconsolately through the streets of London, Uville happened upon the shop in Fitzroy Square. Among the sundry machines and other equipment offered for sale was the working model of Trevithick's high-pressure steam engine. From what Uville had heard of this kind of engine it should work as well on a mountain top as it did at sea level. He paid £20 (£900) for the engine and shipped it back to Peru.

When he arrived in Lima he arranged to have the engine transported to Cerro de Pasco, a mining region high in the Andes. Accessible only on horseback, the 170-mile journey took his small expedition through one of the most terrifyingly precipitous passes in the world. The path was so narrow that the horses and mules had to go in single file.

After many harrowing days he arrived at his destination, ready to put the engine to the test. Lighting a small fire in the boiler the way he had been shown to do in the London shop, he waited anxiously for steam to be raised. Water boiled at a much lower temperature at that altitude – closer to 80° C than 100° C – but there was no shortage of steam. When the pressure was sufficiently high he gave the fly-wheel a nudge. To his profound relief and joy the engine burst into life, working just as well at 15,000 ft in the Andes as it had at sea level in London. Here was the solution to Peru's mining problem, and he resolved to revisit England and search for the inventor of the wonderful machine.

Uville retold his story as his ship approached England. This time his listener was a man named Captain Teague, though his title was neither nautical nor military. Teague was a Cornish mining manager, and he must have been astonished by Uville's accounts of the riches of the Peruvian mines. But the biggest surprise was to be Uville's. 'I know all about it,' said Teague, when Uville told him of his quest to find Trevithick. 'It is the easiest thing in the world. The inventor of your high-pressure steam-engine is a cousin of mine, living within a few miles of Falmouth . . .'. And so began the sea change that would take Trevithick half-way across the world on an eleven-year odyssey into danger, adventure, and undreamed of fortunes.

Trevithick may not have attended the Rainhill trials, but he was in evidence in the machines competing for the prize. The first contender, Braithwaite and Ericsson's *Novelty*, was the least like anything Trevithick had ever conceived, but it did run on high-pressure steam, like all the other mechanically powered entries.

CHAPTER 3

The London challenge

Wednesday 7 October 1829, the second day of the trials, began beneath sullen skies that threatened rain. The wind was blowing mainly from the north, and there was a distinct autumn nip in the air. The large crowds that had flocked to the grounds the previous day were less in evidence, though there was no diminution in the number of gentlemen of science and engineering. Some of these gentlemen may have stopped by *Novelty* to exchange pleasantries with Messrs Braithwaite and Ericsson, but the two were probably too busy to do any more than pass the time of day.

Thirty-two-year-old John Braithwaite still retained his boyish countenance. He wore his hair quite long, as was the fashion, with sideburns reaching down well below his ears. He was slightly built, and his light frame, taken with his youthfulness, gave him an almost foppish appearance. Some might say he looked a typical English gentleman. He was a kindly

John Braithwaite.

John Ericsson.

man, with a sense of humour, and was always considerate to his employees and apprentices.

Ericsson, born and raised in Sweden, looked anything but Nordic with his glossy clusters of thick brown curls. But he did have blue eyes and white skin; he also had a massive forehead. He had a pleasant, easy-going manner and was naturally amiable and generous. But, like a Norse warrior of old, he could fly into an uncontrollable fit of temper when sorely provoked. Weighing twelve stone (168 lb), he was only 18 lb heavier than Braithwaite, but was much more powerfully built, with broad shoulders which he carried with a military bearing. When he arrived in England, just three years earlier, he was still serving in the Swedish army, and although he relinquished his commission the following year when he resigned from the military, he continued using the title of Captain. In spite of their differences in appearance the two men had much in common. Both were skilled draughtsmen, who paid meticulous attention to detail, and both had boundless energy and enthusiasm for their work, coupled with an entrepreneurial interest that drove them to succeed. They were lively conversationalists too, and would have had much to talk about at Rainhill.

Novelty stood at the starting point, already hitched to the wagons that bore her assigned load. All seemed ready, but the two men still had their last-minute checks to make. One of them may have attended to the machinery above the platform while the other checked below. The primary concerns beneath the locomotive were the many linkages between the pistons and the driving axle, and between the eccentrics and the cylinder valves. If any one of the joints should fail *Novelty*'s trial would come to an abrupt end. The longest connecting rods ran almost the full length of the locomotive, and served to transmit power from the paired cylinders at the back, to the axle attached to the driving wheels at the front. Each piston rod was connected to the long connecting rod by a shorter rod, so when the pistons moved up and down they imparted a reciprocating action to the connecting rods. If *Novelty* had been like any other locomotive of her time the far ends of the connecting rods would have been attached direct to the driving wheels. But *Novelty* was unique in having a cranked axle to transmit power from the pistons to the wheels. And here lay the secret of her remarkably smooth motion.

The cranked axle was not original to Braithwaite and Ericsson, but this was the first time it had ever been used on a locomotive. It would eventually become one of the standard drives for locomotives, and is still used today. The device, similar to the crankshaft of an internal combustion engine, is a way of converting reciprocating action into circular motion. Instead of being a simple rod, *Novelty*'s cranked wheel axle had two rectangular half-loops, on opposite sides to one another. So, if the axle were laid flat on the ground, the two loops would lie on either side of the axle's axis of rotation. Each connecting rod was attached, via a bearing, to the bottom of the loop, thereby converting the reciprocation action of the pistons into a rotation of the axle. The cranked axle was an elegant solution to the problem of transmitting power from pistons to wheels without imparting any rolling motion to the locomotive itself.

The other mechanical linkage to be checked was the one that conveyed reciprocating motion from the pistons to the air-compressing system. No detailed description of this system was given at the time, but it was housed in the copper box at the rear of the platform and was a form of flap bellows, used to generate a draught of forced air. The air was delivered to the bottom of the furnace by a pipe, causing the fire inside to burn more intensely. The forced-air system was an integral part of the furnace and boiler. Without this air flow the furnace would be unable to produce enough heat to maintain an adequate supply of steam for the cylinders. As the air pump was mechanically driven, it began working only when the engine was in motion, but it may have been possible to uncouple the drive mechanism and work the bellows by hand, as in the working replica of *Novelty*. Since the rate of pumping increased with speed, the force of the draught, and hence the heat of the furnace and therefore the rate of steam generation, kept pace with the demand – at least in theory.

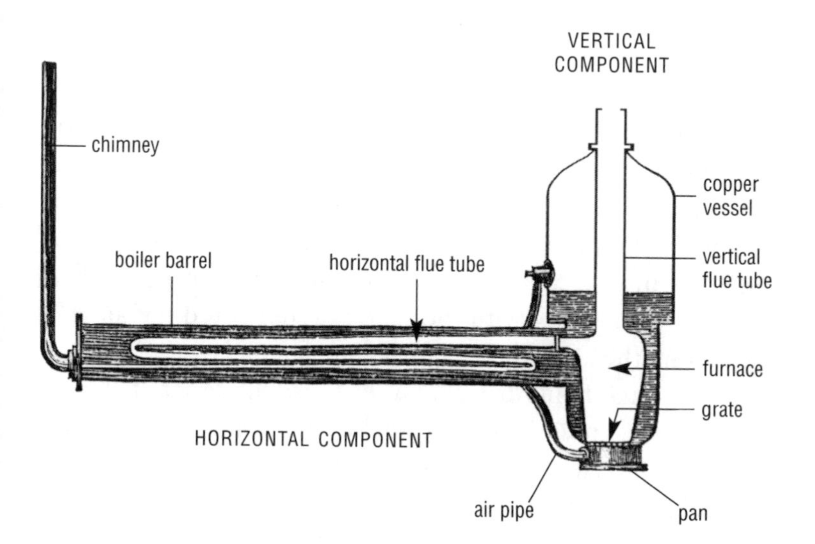

The boiler of Novelty *depicted in the popular Mechanics' Magazine.*

'The boiler of The Novelty is altogether of a peculiar construction,' reported *The Albion* after the first day of the trials, concluding that it 'must involve some new and very valuable principle'. While the reporter for that newspaper was content to leave the secrets of *Novelty*'s boiler as an unopened black box, the *Mechanics' Magazine* gave a detailed description of its workings, illustrated with a simple diagram that explained its basic features quite clearly. The boiler was made in two parts, one vertical, the other horizontal. The vertical portion comprised the prominent copper vessel at the front of the locomotive, which was continued beneath the platform as a slightly narrower chamber housing the furnace. A wide flue tube ran through the centre of the upper chamber, the bottom of which was attached to the top of the furnace. The top of this vertical flue was closed by a tightly fitting lid, which could be slid open for adding fuel to the furnace. The lid was necessary because of the forced-air system: without it, the hot gases from the furnace would simply exit through the top of the wide flue tube, rather than being forced through the horizontal portion of the boiler. Iron bars, appropriately called fire-bars, formed a grate at the bottom of the furnace for the fuel, through which the air draught was forced. The bottom of the furnace was capped with a pan which had the dual role of connecting with the pipe that delivered the air and catching the ashes.

The large boiler barrel of *Sans Pareil*, like that of *Rocket*, was its most prominent feature, but the London locomotive appeared to lack such a structure. The reason for this seeming deficiency was that *Novelty* had a very narrow horizontal boiler barrel – only 13 in. in diameter, compared with just over 4 ft for *Sans Pareil* – and this was largely hidden from view beneath the platform. The boiler barrel was fitted to the furnace, and ran the length of the platform, extending a little way beyond its rear end. Inside the barrel was a long narrow flue tube that looped back upon itself twice.

Although most of the steam was generated in the vertical part of the boiler, the horizontal portion increased the total surface area available for heating the surrounding water. One end of the flue tube was continuous with the pipe that served as the chimney. The other end was jointed, by a flange, to the furnace. The hot gases leaving the furnace therefore looped back and forth inside the flue tube before exiting through the chimney.

The water in the vertical part of the boiler had to remain above the level of the furnace, and this also ensured that the horizontal barrel was always immersed in water too. Heat from the furnace was therefore transferred to the surrounding water through the furnace walls, through the horizontal flue tube, and through the lower part of the vertical flue tube. As the water boiled, steam accumulated in the vertical part of the boiler, raising its pressure. The vertical flue tube, and the furnace, were steamtight, so the lid at the top could be opened for refuelling without any difficulty. High-pressure steam was carried away by a pipe to the cylinders. These had a 6 in. bore, compared with 7 in. in *Sans Pareil* and 8 in. in *Rocket*.

The flanges between the vertical boiler and the furnace and between the furnace and the horizontal flue must have been bolted together, to allow the boiler to be dismantled for cleaning and maintenance. The joints were made steamtight with some kind of cement. During operation the water level in the boiler had to be kept safely above the level of these joints, and above the top of the furnace, to avoid serious heat damage. *Novelty*'s boiler, like that of all other locomotives, therefore had to be continuously topped up with water. Replenishing the water in any high-pressure steam boiler required a feed pump to force the water against the steam pressure, and this had to be fitted with a one-way valve to prevent backflow. Trevithick had realised this three decades earlier, and had fitted most of his engines with a mechanical feed pump, powered by the engine. The same principle was used by all the Rainhill competitors.

Novelty's horizontal flue tube varied in width, becoming progressively narrower from the furnace end towards the chimney. The volume of gases entering the flue tube in a given time was the same as that leaving, because it was a continuous tube with no other outlet. The speed of the flow would therefore have increased continuously towards the narrower chimney end, just as a river flows more rapidly when its banks are narrowed. This was an innovative design because it caused soot and cinders from the furnace to be swept along towards the chimney, rather than accumulating at the other end of the flue where they would have been more difficult to remove.

Although *Novelty* had a smaller boiler than either *Sans Pareil* or *Rocket*, it was hoped that the combination of vertical and horizontal components, and forced-air ventilation, would give it an adequate rate of steam production. Having a small boiler reduced the overall weight of the locomotive, but the saving was not as great as might have been expected from the large difference between the size of its boiler and those of the other two locomotives. *Novelty* tipped the scales at a little over three tons, which was about 72 per cent as heavy as *Rocket* and about 64 per cent of the weight of the much heavier *Sans Pareil*.

Aside from being good business partners, Braithwaite and Ericsson were good friends. They would have thoroughly enjoyed standing shoulder to shoulder at Rainhill, competing for the richest engineering prize in the land. Ericsson could never have found such opportunities in Sweden, but his initial reason for going to Britain had nothing to do with locomotives – at least, not directly. Ericsson's intention was to promote a novel engine he had devised, called the 'flame engine' (also known as the 'caloric engine'). It was designed to use the heat from burning fuel to drive a piston direct, instead of using steam as an intermediary. It was therefore claimed to be more efficient than the steam engine, which he hoped it would eventually replace. Some time in 1825 or 1826 Ericsson sent a manuscript describing his engine to the recently founded Institution of Civil Engineers, in London. He was not the only claimant of a new and better way to generate power, and there would be many more after him, but what set him apart from the others was that he had built a working model of his invention.

Ericsson arrived in England during the summer of 1826 with high hopes of making his fortune, but he was to be disappointed. Sweden had large timber resources, and his engine burned wood, but Britain's industries were driven by coal, and if his engine could not use that for fuel it had no future. Coal burns more intensely than wood, and when Ericsson fuelled his engine with coal the heat destroyed the working parts of his engine. As it happens, the destruction of the flame engine was no great loss because the claim of greater efficiency was based upon unsound science. Ericsson assumed that all the heat energy supplied to the engine would be converted into mechanical energy, but 100 per cent efficiency is impossible. In later years Ericsson would resurrect the idea of the flame engine, building a massive one with 14 ft cylinders to propel a ship, but the project was an expensive failure.

Seeing his dreams disappear in flames was bad enough, but the twenty-two-year-old officer had a more immediate problem to face. He had borrowed money from a fellow officer to finance his trip, and this was rapidly running out.

He desperately needed to earn some money, and his search for employment soon found him a position in London, working for Braithwaite. 'It was my good fortune to meet with Mr Braithwaite's approbation and friendship,' Ericsson wrote many years afterwards. 'In the various mechanical operations we carried out together I gained experience which, but for the confidence and liberality of Mr B., I probably never should have acquired.'

At the time Braithwaite was the sole owner of an engineering works he had inherited from his father. His father, who died in 1818, had made a fortune salvaging the *Earl of Abergavenny*, a barque that foundered off the Dorset coast early in 1805. She was one of the East India Company's largest vessels, and was under the command of John Wordsworth, favourite brother of the fêted poet, the captain going down with his ship in the best maritime tradition. John Braithwaite senior, under contract to the ship's owners, dived on the wreck in a diving bell. Although his son was only nine years old at the time, his father trusted him to watch for his signals from aboard the salvage vessel. The operation lasted two years, during which time he is said to have recovered £130,000 (£6 million).

Ericsson, like countless other enterprising men, had been attracted to Britain because it offered better opportunities than any of the countries on the Continent. France, for example, lacked neither Britain's mineral wealth nor its creativity, and had been the cultural centre of the Western world for generations, but she could not compete in providing such a stimulating entrepreneurial environment. The reason why Britain led the industrial world was primarily because its captains of industry were relatively unimpeded by government bureaucracy. British landowners, unlike their counterparts on the Continent, were free to exploit the mineral wealth of their land, and this motivated businessmen and inventors alike. And, because the richest landowners occupied seats in the Houses of Commons and Lords, they could protect the *laissez-faire* policies that engendered free enterprise. The rich certainly got richer, but there was a trickle-down effect, and the workers who mined their lordships' coal and who worked in their factories had more money in their own pockets as a consequence. Some of this disposable income was spent on manufactured products, like the cheap clothing from the northern mills and the everyday crockery from the Staffordshire potteries, and helped prime the pump of Britain's growing consumer society.

One can only imagine the mounting anxiety aboard *Novelty* as the two men watched the mercury level edging its way towards 50 psi and the start of their ordeal. Going into battle probably felt much the same way, though neither man had experienced combat, not even Ericsson. Since Napoleon's final defeat at Waterloo, fourteen years before, Europe had been at peace. Napoleon himself had been dead eight years, while his nemesis, the Duke of Wellington, was now Britain's Prime Minister. But not even the battle-hardened Wellington would have changed places with the two engineers that morning – he abhorred the very idea of locomotives. The finishing line looked so far away in the distance, and they knew they had to race there and back twenty times before the trial was over. They had already demonstrated they had the fastest locomotive, but how would their unique machine perform in hauling a load of over eleven tons?

Although *Novelty* was a radical departure from all the other locomotives, there were many elements in its design that Braithwaite and Ericsson had used before. The previous year,

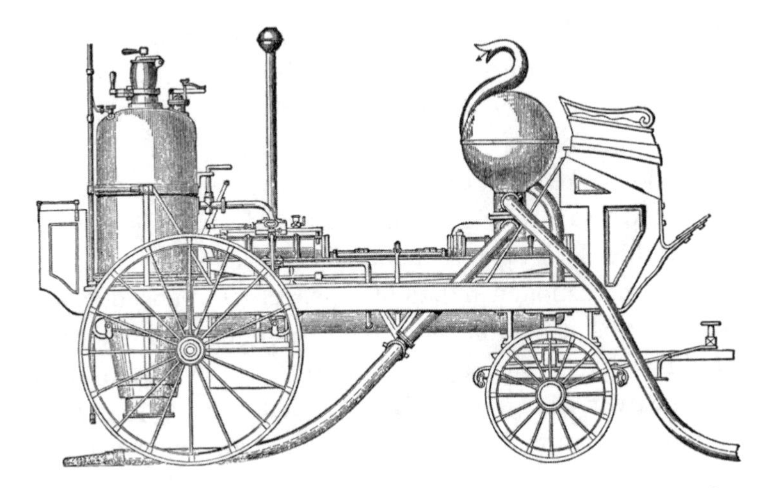

Braithwaite and Ericsson's fire engine was built several months before Novelty and has many similarities with it.

Ericsson had conducted some experiments with a steam-driven water pump for use in fire-fighting. Braithwaite's engineering company had a great deal of experience in pumping equipment, and had also built many steam engines, so bringing the two components together in a new apparatus to fight fires was an obvious development. Early on in the current year they had mounted the equipment on a lightweight frame, with sprung wheels to facilitate bumping up and down the pavement as their machine was pushed to the scene of a fire. Their new fire engine bore a striking resemblance to *Novelty*, the primary difference being that its steam engine was used to power a water pump, rather than driving wheels. The revolutionary engine could shoot a stream of water to the top of the tallest buildings and easily out-performed the other engines, which were all hand-operated. On one particularly cold night, when the other machines had frozen and ceased to function, their engine continued pumping water at a burning building for five hours. In spite of its outstanding success the London Fire Brigade raised a number of frivolous objections against its use, and their superior engine was never adopted. The intransigence to change epitomised by that episode left Messrs Braithwaite and Ericsson feeling totally disgusted. But they made good use of the engineering experience in designing and building *Novelty*.

Any concerns Braithwaite and Ericsson might have entertained regarding the performance of their locomotive were immediately dispelled when the timekeeper's flag fell. *Novelty* dashed away from the starting line, gathering speed smoothly and confidently as she reached out for the hazy speck in the distance. According to the press reports her speed exceeded 20 mph. There were cheers and waves from her enthusiastic supporters standing beside the track, and these grew into a roar as *Novelty* approached the marquee and the thick of the crowd. Hauling her assigned load proved no impediment, and the locomotive 'drew this with ease at the rate of 20¾ miles per hour; thus proving itself to be equally good for speed as for power'. Joseph Locke, an engineer who worked closely with Stephenson, is reported to have said that George Stephenson regarded *Novelty* as the only serious rival to *Rocket*.

Streaking past the mile-and-a-half post they slowed their machine in preparation for reversing directions and making the return leg. If they maintained this speed the prize would surely be theirs – it was unlikely *Rocket* would go any faster, and Hackworth's *Sans Pareil* was hardly built for speed. Timothy Burstall's *Perseverance* was an unknown quantity, but, judging from its battered state, there was some doubt whether it would even be ready to compete. All in all, things looked decidedly promising for Messrs Braithwaite and Ericsson. Landing a contract to supply locomotives to the Liverpool & Manchester Railway would be the making of their company. But that was not the most significant consequence of winning. If *Novelty* triumphed at Rainhill her radical new design would become the standard for all railway engines. Barrel boilers and elephant-trunk chimneys would become a thing of the past as locomotive evolution sped off along an entirely new line.

Novelty came to a halt for the briefest moment, and then she was under way again, gathering speed and heading back towards the starting point. She was running as smoothly as silk and her two attendants were probably beginning to feel in their stride too. Everything was going just perfectly.

Suddenly there was a dreadful noise from the copper box at the rear of the platform. The bellows had suffered a catastrophic failure. Robbed of its vital flow of air, the fire in the furnace died down, steam pressure fell and *Novelty* slowed down. As the stunned crowd looked on, hardly believing their eyes, their *beau-idéal* coasted to a halt. George Stephenson is said to have turned to Joseph Locke at this point, commenting in his broad Northumbrian accent, 'We needn't fear yon thing; it's got no goots.' Braithwaite and Ericsson probably realised what the problem was directly they heard the noise, and it would have taken them little time to assess the extent of the damage. It was a major mechanical failure, one that could not be repaired in a few hours. After conferring with the judges a decision was made: the trial of *Novelty* would be postponed, allowing time to repair the machine.

The crowd needed something special to make up for their disappointment, so *Rocket* was rolled out, steam was raised, and a coach hitched capable of carrying thirty passengers. And while *Rocket* streaked up and down the track, electrifying the brave souls aboard as she reached the unbelievable speed of 30 mph, the judges paid a visit to *Sans Pareil.* Timothy Hackworth was still working away to seal the leaks in his imperfect boiler. Nevertheless, at the judges' behest, he put down his tools and began preparing his engine for running on the track. But as the steam pressure began to rise, it became abundantly clear that the boiler was still leaking too badly to be able to continue. The judges therefore granted him more time to complete his repairs. Timothy Burstall's battered *Perseverance* was even further from being ready, so Messrs Wood, Rastrick and Kennedy directed Stephenson to present *Rocket* for trial first thing in the morning.

Thomas Brandreth was no doubt delighted when he was invited to give another demonstration of *Cycloped* for the entertainment and edification of the crowd. His horse-powered machine could hardly compete with the dash and thrill of a steam locomotive, but it did at least make a change from *Rocket,* which was running the risk of monopolising the entire event. Just how long *Cycloped* performed before the accident occurred is not known, but several newspapers, including *The Times*, reported the unfortunate mishap that ended the demonstration:

> One of the horses engaged in working Mr Brandreth's carriage fell through, we understand, the floor, but was extricated without much injury. [The account went on to describe the deterioration in the day.] The weather became wet, and the rail-ways clogged with mud, which made it necessary to suspend the prosecution of the experiments before the day had half elapsed.

The weather had become quite atrocious, and thousands of milling feet transformed the ground into a quagmire.

Some accounts gave the impression that *Novelty* had to stop because of the wet conditions. Included among these was that in the *Mechanics' Magazine*, which made no mention of the bellows failure. Having reported the ease with which *Novelty* had hauled its assigned load, it stated that:

> We took particular notice to-day of its power of consuming its own smoke, and did not any time observe the emissions of the smallest particle from the chimney.
>
> The weather now became wet, and the rail-ways clogged with mud, which made it necessary to suspend the prosecution of the experiments before the day had half elapsed.

The passage referring to the suspension of the experiments is identical to the one used in *The Times*. The exact same wording also appeared elsewhere, including in *The Albion*, showing they derived from the same source. There are similar occurrences of identical passages in many other Rainhill articles, revealing that there were fewer independent accounts of the trials than was apparent. One of the contributors was Charles Vignoles (1793–1875), a young engineer with an active interest in railways, who reported on the trials for the *Mechanics' Magazine*. Vignoles had a great deal of admiration for Messrs Braithwaite and Ericsson, and for their revolutionary new locomotive. He also had a deep-seated dislike of George Stephenson, with whom he had crossed swords. But Vignoles was only one among many who had run afoul of George Stephenson.

As the rain lashed down and the bedraggled spectators picked their muddy way from the grounds, Braithwaite and Ericsson contemplated their plight. *Novelty*'s air-compressing apparatus was extensively damaged and it would take some time to make the necessary repairs.

Hackworth's situation was more serious. The repairs to *Sans Pareil*'s boiler were a major task, and had to be completed before he could raise steam and carry out the myriad adjustments and fine tuning in preparation for his trial. And there was so little time.

CHAPTER 4

A man of principles

The sharp clash of hammer against cold chisel would have been a familiar sound during the early days of the trials, probably ringing out into the night long after the last visitors had left. The deft hands wielding the tools drove the narrow threads of copper deep into the cracks between the riveted wrought iron plates, making the ductile metal yield and flow.

Having one or two small leaks in a boiler was neither unusual nor troublesome, and those in the water-filled lower section eventually sealed themselves anyway, simply by rusting as the water seeped out. But it was a slow process, because wrought iron is so resistant to corrosion. Leaks in the steam-filled upper portion were much harder to stem because steam, being a gas, passes right through the finest gaps. The leaks in *Sans Pareil*'s boiler were so extensive that Hackworth had no alternative but to cold-caulk the seams with copper, a slow and laborious process.

Riveting was the only method of joining the boiler plates together, and this was necessary because wrought iron was available only in relatively small pieces. Even if gas or arc welding had been available at the time of Rainhill, it could not have been used because wrought iron is almost impossible to weld. Not that riveting is inferior to welding, and in some applications, like ships' hulls and large boilers, it is actually superior because of the limited amount of movement riveted joints permit.

The skill of the boilermaker in cutting, shaping and fitting the individual plates had to be matched by the competence of the riveter in joining them together. The rivet holes had to be perfectly aligned, and the overlapping edges of the plates pressed firmly together in tight contact. And as he drove each red-hot rivet home, he had to make sure his assistant held the head end firmly while he hammered the other end, spreading it like a mushroom. As the hot rivet cooled and contracted it locked the plates tightly together without any gaps.

If Timothy Hackworth had overseen the building of the boiler there would have been no leaks at all because he would have made sure the boilermaker did his job properly. But he had enjoyed neither the time nor the opportunity to visit the Bedlington Iron Works; he had just had to trust the contractor to do a good job. It was not that the ironworks had a bad reputation for building boilers, but riveting was a highly skilled trade and there was a wide range of talent among its craftsmen.

There was also great variation in the quality of the wrought iron used, and in how well it was rolled into plates. At that time wrought iron could be made and rolled only in small quantities, which explains why boiler plates were of limited size. It also explains why wrought iron was more expensive than cast iron. The major difference between the two types is that wrought iron is more refined, reducing the impurities, and is considerably stronger, in part because it has been rolled or beaten with hammers. Aside from its superior strength and toughness, wrought iron does not suffer from the brittleness of cast iron. It is also malleable, and can be shaped by hammering, whereas cast iron would be shattered by such treatment. Although the brittleness of cast iron makes it unsuitable for many applications, including rails, it is an excellent casting material and was extensively used for making cylinders, wheels and wheel bearings; it is still used today for making engine blocks and crankcases.

The quality of wrought-iron plate varied in the uniformity of its thickness. George Stephenson was quite critical of the Bedlington Iron Works on this last point. The year before Rainhill he warned his son Robert that he 'must take care that you have no thick plates and thin ones, as is often the case with those which come from Bedlington'. His advice was duly taken, and *Rocket*'s boiler was constructed from the 'Best RB' rolled wrought-iron plate, obtained from a different supplier.

Hackworth's frustration over the leaking boiler was probably exacerbated by the knowledge that Michael Longridge, of the Bedlington Iron Works, was one of George Stephenson's business partners. Frustration was accompanied by anger, and Hackworth had been in ill humour since his arrival at Rainhill. John Dixon, a resident engineer on the Liverpool & Manchester Railway, wrote that 'we could not please him with the Tender or anything; he openly accused G.S.'s [George Stephenson's] people of conspiring against him, of which I do believe them innocent . . .'.

While it is unlikely that anyone in the Stephenson camp did anything to hinder Hackworth's chances of winning at Rainhill, he had good reason to feel the odds were stacked against him. Hackworth had to build *Sans Pareil* entirely in his own time and at his own expense, using the last few hundred pounds of a small inheritance to complete the job. The work was carried out in the Stockton & Darlington's cramped and poorly equipped repair shop at Shildon. The demands of his regular job meant he had to work long into the night to complete the task, and he had to contract out all the major components, not just the boiler. Stephenson, in contrast, had unlimited means, and a fully equipped and staffed locomotive factory of his own, in Newcastle. He also had Edward Pease, a wealthy businessman who was also one of the most influential board members of the Stockton & Darlington Railway, as one of his business partners. With all these resources available, *Rocket* was finished in plenty of time to allow for track testing and final adjustments. Poor Hackworth had no such opportunity, and his engine barely arrived in time for the start of the trials.

Timothy Hackworth.

The water colour of Timothy Hackworth, painted some time during his middle years, portrays a rather dour-looking figure. Even without prior knowledge of his devotion to Methodism, his expression could still be interpreted as sanctimonious. Some measure of the man can be gleaned from the somewhat punctilious letter he wrote to Edward Pease at the time he was being considered for a job. To put the letter into context, Pease was a man of enormous influence. During the early days of the Stockton & Darlington he had singlehanded saved it from oblivion with a personal subscription of £10,000 (about £500,000). Pease had recently interviewed Hackworth for the position of superintendent.

> Dear Sir,
>
> I have waited rather impatiently to know your determination as to my proposed engagement with you, and not having been favoured with your decision on the subject, I am just on the point of fixing on a situation to commence business on my own account at Newcastle: but I feel it to be my duty to give you this intimation of my intention . . . If I should not be favoured with your reply in the course of the two next posts, I shall consider myself fully at liberty . . . I cannot submit to a less salery [*sic*] than I proposed.
>
> Yours truly
>
> Walbottle May 9, 1825 Timothy Hackworth

Hackworth came highly recommended for the position. The owner of the Walbottle Colliery, where he was employed as foreman smith, wrote that he was 'an Ingenious, Honest,

Industrious man [who] has served us faithfully for eleven years . . . He is sensible, and truly respectable . . .'. Pease probably already knew of Hackworth's qualities through his discussions with George Stephenson. Hackworth was a talented blacksmith, skilled in metalwork and in boilermaking, and was just the kind of all-round practical man the company was looking for. But Pease, a straitlaced Quaker, must have been equally impressed by his moral rectitude. When Hackworth had been asked to work on a Sunday at a previous job he had preferred resigning to compromising the sanctity of the Sabbath.

Hackworth got the job, so becoming the first superintendent of the first public railway in the world. He was responsible for the day-to-day operation of all the locomotives and the entire track, which included the inclined planes with all their rope haulage equipment. Such heavy responsibilities should have been financially rewarding, but he received an annual salary of only £150 (£9,000). From this he had to pay his employers £12 (£600) a year for renting one of their houses, and £4 4*s* (£200) for the coal to heat it. Drivers were earning about twice as much, for considerably less responsibility and shorter hours.

Hackworth had learned his trade in the time-honoured tradition of serving a seven-year apprenticeship. Few families could afford indenturing their young sons in this way, but his father had a well-paying job – he was foreman blacksmith at the nearby Wylam Colliery – and realised his son's 'natural bent and aptitude of mind for mechanical construction and research . . .'. To give his son the best possible start in life he had also kept him at school until he was fourteen. This was remarkably progressive for the early 1800s because there was no state education until 1870, and it was not until 1914 that the school-leaving age was raised to fourteen. Most of the boys young Hackworth knew would have been sent out to work long before that age, and many would have been working in the coal mines from as early an age as six. Hackworth started his apprenticeship in 1801, working under his father at Wylam Colliery. Christopher Blackett, the owner, was as impressed by the son as he was by the father, and took a particular interest in his progress. When Hackworth senior died the following year, Blackett pledged he would have his father's old job when completed his apprenticeship.

Timothy Hackworth's promotion to foreman blacksmith was most timely, because it coincided with Blackett's renewed interest in steam locomotion. For some time now he had wanted to run a steam locomotive on the five-mile wagonway linking his colliery with the shipping wharf. Quite aside from any academic interest he may have had in doing this, there was a sound economic one. The continuation of the Napoleonic Wars had significantly inflated the prices of hay and grain, adding substantially to the cost of hauling coal by horse wagon. Finding a cheaper alternative to using horses was therefore in every colliery owner's interest. But before Blackett could get started he would have to replace the wooden rails with iron ones; otherwise a heavy locomotive would demolish the track.

Wagonways had been used since the early 1600s for hauling heavy loads over short distances, either by horse or by hand. Hauling carts and wagons along a wagonway was considerably easier than bumping them over rough ground, and laying down two parallel tracks was much cheaper than building an entire width of roadway. As trees were fairly plentiful wood was used, but it is not very hard-wearing, especially when heavy horse carts were trundled back and forth, so the track had to be continually replaced. Iron wears well, but was too expensive. However, the advent of coke-fired blast furnaces during the early eighteenth century considerably reduced the cost of cast iron, making it more readily available. Laying iron wagonways therefore became economically feasible, and the wooden ones were gradually replaced.

The iron rails came in a variety of shapes and sizes. The simplest ones were merely flat oblong plates. Others were L-shaped in cross-section, the vertical flange being used to guide the wheels, thereby keeping the vehicle on the track. Both kinds were known as *plate* rails. An alternative way of preventing derailment was to place the flange on the edge of the wheel instead of the rail. The flat part of the wheel ran along the top edge of the rail, and this was rectangular rather than flat in profile. This type of rail, still used today, is called an *edge* rail. Plate rails were especially popular in Wales, and the Pen-y-darren railway, where Trevithick ran his first locomotive, was of this kind. Perhaps that is why Blackett replaced his wooden track with plate rails. Be that as it may, when Trevithick declined Blackett's request for assistance in building a locomotive, the colliery owner knew he would have to make his own arrangements.

While Blackett 'knew nothing of engineering' himself, he did have some very capable men working for him. Foremost among them was William Hedley, his viewer, who took charge of the locomotive project. All collieries had a viewer, a position usually interpreted as that of mining engineer, or superintendent. Viewers were usually skilled in all things practical, and many of them went on to become eminent in their own right. Nicholas Wood, for example, one of the three Rainhill judges, had been a viewer at Killingworth Colliery, Northumberland, and Blenkinsop had been a viewer at Middleton Colliery in Yorkshire. Before proceeding to build a steam locomotive, Hedley's first task was to make absolutely sure there would be sufficient traction between a locomotive's smooth wheels and the smooth iron track to enable it to haul an adequate load. Slippage between the two smooth surfaces had long been perceived as a potential problem for locomotives, and Trevithick had proposed a solution back in 1802. To make his patent for a road carriage more comprehensive, he had referred to railroads, and briefly mentioned improving traction by studding the wheels with small projections. As it happened this proved unnecessary, and his Pen-y-darren locomotive had smooth wheels. But it seems likely that neither Blackett nor Hedley knew this, because Hedley devised and built a full-size test carriage, probably in 1812, to assess the amount of traction that could be generated under various loading conditions. Mounted on two pairs of chest-high wheels, it was powered by a team of men turning handles that were geared to the wheels by a series of cogs. Hackworth undoubtedly helped with its construction, and probably participated in the experiments too.

At the time that this experimental work was proceeding at Wylam, a novel locomotive was being tested at Middleton Colliery, some sixty miles to the south. Its unique feature was that it was propelled by a driven cogwheel on one side of the engine, between the front and back wheels, which engaged with a notched rack on that side of the rail, thereby eliminating the traction problem altogether. Blenkinsop is usually given sole credit for the invention, but it was a co-operative venture with the manufacturer, Fenton Murray & Wood, of Leeds, and the locomotive design was the work of Matthew Murray. The locomotive had twin cylinders, eliminating the need for a flywheel, which was a marked improvement on Trevithick's

Pen-y-darren and *Catch-me-who-can* locomotives. In contrast to these, though, it lacked a feed pump for the water, so the locomotive had to be stopped periodically to refill its boiler.

The railway experiments conducted at Wylam, Middleton and Pen-y-darren show that the results were not freely shared. If those involved had been more communicative, and fully aware of what others were doing in the field, they could have built upon their collective experience. The early evolution of the locomotive might then have followed a more linear progression, rather than its meandering pathway with all the stops and starts and blind alleys.

Blenkinsop resolved the traction problem by using a cogged wheel drive.

Blenkinsop's locomotive was demonstrated before a large gathering of interested spectators, an event reported in the local *The Leeds Mercury* on 27 June 1812. At one point it hauled 'eight wagons of coal, each weighing 3¼ tons . . . to which, as it approached the town of Leeds was superadded about 50 passengers . . .'. It completed the journey between the colliery and the shipping wharf 'without the slightest accident', at a little over three miles an hour.

Blenkinsop claimed his locomotive could haul a train of coal wagons weighing over 100 tons, and that it was unimpeded even by the heaviest snowfall. He also promised it would save 'five-sixths of the expense of conveying goods by horses . . .'. Given the high cost of feed, the last claim struck a chord with mine owners, and three other collieries besides Middleton adopted the system. This was the steam locomotive's first commercial success but it was a cumbersome system, with high wear and tear on the rack-and-pinion mechanism, and required regular maintenance. By the time of the Rainhill trials Blenkinsop's system was much less used, though it was still operating at Middleton.

A more creative solution to the imagined traction problem was offered by William Brunton, with a locomotive propelled by a pair of mechanical legs. This bizarre contraption, named the 'Mechanical Traveller', moved along at 2–3 mph, but it was not successful. It blew up in its second year, killing several people and causing extensive property damage.

The Wylam experiments removed any lingering doubts regarding traction, and Blackett was keen to start building his first locomotive, but finding an experienced man to undertake the task was a challenge. After some searching a suitable engineer was found in Thomas Waters, who owned an engineering works in Gateshead, near Newcastle. Blackett probably engaged him early in 1813, and work began right away. The year 1813 was also the one in which the twenty-seven-year-old Hackworth was married. His bride, like himself, had broken from the Anglican Church – that omnipotent bastion of British morality that dominated all aspects of life – to become a Methodist. It was no small step to take in those days, and she had been forced to leave home and live with a relative.

The locomotive was built on the test carriage, and completed in about three and a half months. It had a cast-iron boiler with a straight-through flue, and a single cylinder coupled to a flywheel. The steam pressure peaked at about 50 psi, but the boiler had such a poor steaming capacity that the pressure could not be maintained. The locomotive was accordingly sluggish and temperamental, often running out of steam and coming to a complete stop. When it did so during its first test run Waters, who was driving, lashed down the safety valve and instructed his fireman to stoke up the fire. The onlookers, fearing the worst, retreated to a safe distance, but there was no explosion, and the locomotive completed its trial without mishap.

The locomotive was not the success Blackett and Hedley had hoped for and the temptation to cut his losses and abandon the project might have crossed Blackett's mind. But he was determined to carry on, and proceeded with a second locomotive. Waters does not appear to have been involved this time, and the engine was built almost entirely under Hedley's supervision. To increase the heating area, thereby improving the steaming capacity, a return-flue boiler was built. This was more difficult to construct than a straight-through flue and a skilled craftsman was hired for the task. Hackworth's talents as foreman blacksmith would have been invaluable, and his involvement in building this type of boiler influenced him for the rest of his career.

The second locomotive was a considerable improvement upon the first, with a reliable boiler that could maintain a good head of steam. It could haul a train of sixteen loaded coal wagons at 5 mph, thereby doing the work of more than a dozen horses. And, since it burnt the lowest-quality coal, fuel costs were minimal. The large work-force Blackett employed for managing the horse-drawn traffic could see the threat to their livelihoods, and were openly opposed to the experiments. They took great delight every time the locomotive broke down and had to be ignominiously hauled back to the workshop by horses, and if the infernal machine had blown itself to Kingdom Come they would have been overjoyed. Far from being destroyed in a boiler explosion, though, this locomotive appears to have survived to the present day.

London's Science Museum, tucked unobtrusively around the corner from the Natural History Museum, houses a treasury of relics from the industrial revolution. The ground-floor gallery, 'Making the Modern World', is the final resting place of *Rocket*'s remains, and a few yards further on is *Puffing Billy*. Built some time around 1815, it is the oldest surviving locomotive in the world. Little documentation has survived, so we cannot be sure, but *Puffing Billy*, although perhaps rebuilt, may be the second Wylam locomotive. And, if it is not, it must be very similar because several locomotives of its type were built during this period, all closely similar.

Standing beside the almost 200-year-old locomotive engenders an enormous respect for the ingenuity and industry of those who built it, and this includes Hackworth. Imagine riveting together all those heavy wrought-iron plates to make a boiler that did not leak. The back end of the boiler, the one furthest from the furnace and chimney, is domed into a perfect hemisphere. A dozen petals of iron form the periphery of the dome, each overlapping its neighbour along one edge, only to be overlapped itself by the next curved plate. A dished circular plate forms the centre of the dome, overlapped around its circumference by the other twelve plates. The iron florescence reminds me of nothing so much as the ring of bony plates found in the eyes of prehistoric reptiles, on display at the Natural History Museum. The man who riveted the boiler was a master craftsman and his legacy is a work of art.

A huge pair of cylinders is mounted vertically on either side of the boiler, in line with the rear wheels, with the piston rods directed upward. They are made of riveted plates too, and their contact with the boiler is so intimate that it is difficult to see where one ends and the other begins. The tall chimney, and the U-shaped flue tube inside the boiler barrel, are also riveted, but their plates are thinner. Above the boiler is a confusing array of rods and levers. Prominent among them is a pair of wooden beams, running the length of the boiler. Each is pivoted at the front to an iron support anchored to the top of the boiler. The free end is attached to the piston rod of that side. The movements of the piston caused the beam to rock up and down, reminiscent of an atmospheric engine. A long connecting rod, attached about mid-beam, transmitted power to the wheels on both sides, through pedal-like cranks and a series of intermeshing cogs. An additional connecting rod was attached to the right-hand beam to provide power for a feed pump, used for forcing water into the boiler against the steam pressure. Exhaust steam from the cylinders was discharged into the chimney, improving the draught through the furnace. The Wylam locomotives were a modest success, setting the pattern of locomotive design for the next fifteen years.

Wylam received many casual visitors during its early days of steam locomotion, people often turning up out of idle curiosity. Young George Stephenson was a regular visitor. He worked at Killingworth Colliery, and used to walk some twelve miles to Wylam after work most Saturday afternoons, often accompanied by Nicholas Wood. They also used to visit nearby Coxlodge Colliery, where a Blenkinsop engine was at work. Visitors to Wylam were no problem at first, but things changed when Blackett learned that Sir Thomas Liddell was planning to build a locomotive at Killingworth, with his employee, George Stephenson, in charge of the project.

Sir Thomas Liddell, later Lord Ravensworth, was one of the Grand Allies, a triumvirate who owned or otherwise controlled most of the collieries in the region. The other two were the Earl of Strathmore and Mr Stuart Wortley, later Lord Wharncliffe. The insatiable demand for coal to power the burgeoning factories and heat the workers' homes had made all three men very rich. London, housing 10 per cent of the country's population and spreading like the pox, was their largest market, and the more smoke that spewed from the chimney pots the more money filled their coffers. Blackett knew that if the Grand Allies were interested in steam locomotion there would be no stopping them, and since he had no wish to see them profit from his own efforts, visitors to Wylam to see the locomotives were no longer welcome.

Hackworth had a good life at Wylam. The place of his birth had given him a good living from a job he enjoyed, working for an appreciative employer in the exciting new field of steam locomotion. His views on keeping the Lord's Day were well known and had always been respected, yet one day he was asked to do a job at the colliery on a Sunday. His strict religious principles forbade it – besides, he had his preaching and visiting the sick on the Sabbath – and he politely refused. The incident prompted him to seek employment elsewhere. Hackworth enjoyed a good reputation, both for his skills as a craftsman and for his good character, and it was not long before he was offered the position of foreman smith at Walbottle Colliery, some three miles away. There were no locomotives at Walbottle, and no plans to build any, but Hackworth knew he would never again be asked to work on a Sunday. He accepted the offer and took up his new post late in 1815 or early in 1816.

While Hackworth was settling into his new job and new home in Walbottle, Trevithick was being lionised in Lima. He had just disembarked, accompanied by a large quantity of machinery, and his arrival was reported in the local press in glowing terms. He was described as 'an eminent professor of mechanics' whose 'arrival in this kingdom will form the epoch of its prosperity . . .'.

Trevithick had left Penzance four months earlier aboard a South Seas whaler, the only vessel available for his passage to South America. Uville, accompanied by three Cornishmen, had left two years earlier, with the first consignment of machinery, totalling £16,152 1*s* 1*d* (almost £700,000). Because of Uville's cash-flow problems, Trevithick had contributed some £3,000 (£128,000) from his own pocket towards this cost. Furthermore, instead of being paid for his considerable time and expertise in selecting and acquiring the equipment, he had accepted shares in the mining interests Uville represented.

That Trevithick willingly invested so much of his own money and time on the promises of a man he hardly knew speaks volumes. Trevithick was a romantic, a dreamer, a man whose gambling instinct overrode any business sense he might have had. One wonders if he would have changed his plans had he known that Uville had signed a partnership agreement before leaving Peru, strictly forbidding additional partners. Uville had also spent more than twice as much money on equipment than had been authorised. He was not an honest man.

Hackworth's years at Walbottle may have been the happiest of his life. He fitted into a steady, if unchallenging, work routine which left him with time and energy to enjoy his family, his leisure activities and his service to God. He was a contented man, a man with talent but little ambition. Hackworth had none of the passion that drove Trevithick from one new project to the next. Nor did he have George Stephenson's pit-bull determination to win against all odds. Even his employer said 'that his great talents – buried there in obscurity – fitted him for much higher and more important duties . . .'. He might have stayed on at Walbottle for the rest of his life, but was persuaded to go and work for George Stephenson at the locomotive works in Newcastle. Finding this not to his liking, he soon left, to try his hand at business. This did not last very long, either, and he took up the position of superintendent of the Stockton & Darlington Railway in May 1825, five months before the railway began operations.

The contrast between the tranquillity of his former job at the colliery and the turmoil of the railway must have been absolute. There were three systems in operation along the track: steam locomotion, horse traffic and rope haulage. Having charge of any one of them would have been work enough for any man, but Hackworth was responsible for all three. And the one that gave him most trouble, right from the very start, was the rope haulage system, which he had to redesign. It would never be perfect, but was much improved for Hackworth's attention. He put his sound practical knowledge to good use in resolving many more of the everyday problems that arose on the world's first public railway. But his most significant contribution during his early years, and certainly the most important one in preparing him for Rainhill, was the building of a locomotive named *Royal George*.

The shortage of serviceable locomotives was one of the railway's major problems, and this became acute in the autumn of 1827 when *Hope*, one of only five, received a 'grievous injury' in a crash. Hackworth proposed building a new locomotive in the company's own repair shop at Shildon. The directors approved, but stipulated that it had to be built as cheaply as possible. To reduce costs it was decided to dismantle the ineffectual *Chittaprat* – named for the noise it made in venting steam – and salvage its boiler, replacing the original flue tube with a return flue, made of wrought iron. Hackworth's design for the new locomotive was radically different from all others of its time. Instead of mounting the vertical cylinders in the traditional way, set in the top of the boiler and thrusting upward, Hackworth mounted them on the outside of the boiler, at the rear, and turned them through 180°, so they fired downward. Each piston rod was now attached directly to its driving wheel, by a small crankpin, eliminating the overhead rocking beams and all the associated brackets and rods. In hindsight this more direct way of transmitting power to the wheels seems obvious, but it took Hackworth, building a locomotive at minimum cost in a small repair shop, to do it. *Royal George* was larger and heavier than Stephenson's *Locomotion* – the first engine in service on the Stockton & Darlington Railway – which is why Hackworth mounted it on three pairs of wheels rather than two. Like the other locomotives on the line, the wheels were connected on either side by a coupling rod. The first two pairs of wheels were sprung, but there were no springs on the driving wheels at the back because they would have worked against the vertical thrust of the pistons, absorbing some of their force. As in the Wylam locomotives, the exhaust steam was ducted to the chimney to assist the furnace. But some of the steam was piped to an iron tank connected to the water tank in the tender. By adjusting the steam supply, the driver could pre-heat the water before it was pumped into the boiler. Other innovations included an improved and adjustable spring-loaded safety valve, and self-lubricating bearings on the wheels.

Hackworth's Royal George, *completed in 1827, was the most advanced locomotive of its time.*

According to James Walker, a contemporary engineer of some note, *Royal George* was 'undoubtedly the most powerful [locomotive] that has yet been made . . .'. George Stephenson's youngest brother, John, happened to be visiting his brother James at the time *Royal George* was undergoing its trials. He was so impressed with what he saw that he declared it to be 'the finest in the world . . .'. He returned home full of praise for the revolutionary new engine. George Stephenson would have listened with great interest.

CHAPTER 5

Up from the mine

George Stephenson would have been hard to miss at Rainhill with his outgoing manner and friendly demeanour. This was his part of the country and his kind of people: enginemen, colliers, pit owners; men with dirt under their fingernails and brass in their pockets. And this was his kind of business: iron rails, stone bridges and steam locomotives, the nuts and bolts and brawn of the transport revolution he helped forge. He would have known many of the men who were there – at least the northerners – and almost all of them would have known of him, even if their paths had never crossed. George Stephenson had become something of a northern legend. There is little doubt that without his pioneering work during the embryonic days of the locomotive, and his influence during the period leading up to the trials, Rainhill would never have taken place.

Stephenson had entered *Rocket* for the Rainhill trials with his son Robert and Henry Booth. He was therefore an integral part of the contest, though he probably left the organising to Robert, and there were plenty of others to

George Stephenson.

take care of the details. He could therefore perambulate the grounds almost as a spectator, at least when he was not stealing the limelight aboard *Rocket*.

He would have thoroughly enjoyed doing the rounds, renewing an old acquaintance here, having a chat there. He was one of them, and even though he now moved in higher circles, consorting with the rich and powerful captains of the new industrial age, he never forgot his old friends. And he still spoke with the same thick north-country accent that southerners found almost impossible to understand. Almost five years before, during his gruelling appearance at the parliamentary hearings on the Liverpool & Manchester Railway Bill of 1824/5, his heavy accent had been a source of derision. But derision had turned to debacle when the parliamentary committee found errors in the survey of the line, which had been carried out by his junior surveyors. The Bill was ultimately defeated, primarily owing to the objections of the local landowners, but the surveying errors had not helped the company's cause. Determined to succeed the next time, the directors hired the celebrated London engineers George and John Rennie as consultants, over Stephenson's head, which did not sit well with his enormous pride. A new survey was ordered, avoiding the estates of the most vigorous opponents, conducted by Charles Vignoles, working under the direction of the Rennies. This explains Stephenson's hostility towards Vignoles, and his disdain for the London engineers, with their institutional elitism. The establishment of professional engineers had treated him with contempt from the start of his career. As far as they were concerned he was little more than an impostor. He lacked their theoretical knowledge and their professional qualifications, shortcomings that could never be replaced by native wit and guile. As for not having gone to the right schools, he never went to *any* school.

In the home that George Stephenson grew up in there was no money for sending children to school. His father had a good steady job at the pithead, working as a fireman of an atmospheric engine, but it only paid 12*s* a week (£55), which was not a large wage even back in 1781, when George was born. His parents and their six children lived in a single room of a four-room labourers' house, a 'two up and two down', shared with three other families. The colliery owned the house and the rent was part of his wages, so there were no accommodation expenses. Even so, 12*s* did not stretch very far when feeding and clothing eight people. Although the Stephensons were poor, like most of their neighbours, they were happy. Robert, his father, was a most congenial soul and 'Bob's engine-fire' was a popular evening rendezvous for the local youngsters. There they would sit in the glow of the engine's furnace while he beguiled them with tales of Sinbad the sailor, Robinson Crusoe and stories from his own fertile imagination.

Like all the other families who owed their living to the mine, the Stephensons were always on the move, following new pits as they opened like birds behind the plough. Leaving Wylam, they moved to a one-room cottage at Dewley Burn, where the Duke of Northumberland had just opened a new pit. George and his three brothers had to get jobs as soon as they were old enough to contribute to the family income, but his two sisters stayed at home to help their mother with cleaning and washing, and preparing meals. It was at about at this time that George, now eight, got his first job, looking after a small herd of cows for a neighbouring widow. A wooden wagonway ran along in front of the Stephensons' home, used for hauling chaldron wagons of coal from the colliery. The widow had grazing rights along the track, and young George's job was to keep the cows from harm's way when the horse-drawn wagons trundled by. In years to come the wooden rails would be replaced by cast-iron ones, making way for locomotives. Meanwhile the countryside was peacefully quiet, and George could while away his two-pence-a-day time bird-nesting, digging for clay and making reed whistles.

However happy young George may have been, he yearned to have a proper grown-up job at the colliery like his father and his older brother James. He did not have long to wait. His first employment at the colliery was as a 'corf-bitter', the same as his brother. His job was to pick stones and other impurities from the coal, for which he was paid 6*d* a day. This was raised to 8*d* when he graduated to driving a horse-whim, and then to 1*s* (about £5 per week) when he was appointed assistant fireman to his father. George was now fourteen and well on his way to manhood.

When the coal ran out at Dewley Burn the family moved on to another of the duke's coal mines, just a few miles away. By this time all the boys were working in the colliery, giving the Stephensons a combined weekly income of between 35*s* and £2 (£105–£120) a week. By comparison with other families they were quite comfortably situated, but these were inflationary times, and food prices, especially that of bread, were astronomically high. Even before the outbreak of the Napoleonic Wars in 1793, when wheat and other grain could still be imported from the Continent, prices were outrageously high. Bread prices during the six years leading up to the war averaged about 7*d* a loaf (over £2) and it is little wonder there were bread riots throughout the land. But these prices doubled during the twelve years of war. Coal mining was an essential industry and the colliery workers were among the best paid of the working classes, especially during the war. How the less fortunate families managed to survive is hard to imagine. Many simply did not survive, and over 10 per cent of the population were reduced to pauperism, entirely dependent on the meagre charity grudgingly handed out by the parishes.

George Stephenson's next promotion was to engine fireman, with an assistant of his own, and it was not long after this that his wages increased to 12*s* a week. This was as much as his father was earning and marked an important point in his life. When he left the foreman's office with his new pay packet he proudly announced to his workmates, 'I am now a made man for life!'

Stephenson was very competitive and always enjoyed taking up the challenge of his workmates in feats of strength. He was a wiry fifteen-year-old, but was deceptively strong, and it was said he had no equal in tossing the hammer. One of his principle challengers in these feats of strength was Robert Hawthorn, several years his senior. Hawthorn subsequently moved on to take up the position of engineer at Walbottle Colliery, where Hackworth would work almost two decades later. His first task was to erect an atmospheric pumping engine, and it was probably through his recommendation that George got the job as its engineman when it was completed. George was only about seventeen at the time, which was very young for such a responsible job. His father went along as his fireman, and although he was now a rung lower on the colliery ladder than his son, old Robert Stephenson was probably as proud of George as George was of himself.

Pay-day at the colliery was on alternate Saturday afternoons, at the end of the working day. Most of the men then went off drinking at one of the local inns, or amused themselves waging money on the dogfights and cockfights in the fields beside the colliery. But George Stephenson never took part in any of these pastimes. His passion in life was finding out all he possibly could about how things worked, and he spent his Saturday afternoons dismantling and cleaning his engine. It was a labour of love, and he spent countless hours totally immersed in valves and pipes and pistons. His engine was his pride and joy, and he wanted to know exactly how it was built and how it worked. He was an astute observer, with an exceptional ability for things mechanical. He became so familiar and competent with the mammoth engine that he could fix almost anything that ever went wrong, rarely having to seek the help of the chief engineer. His natural curiosity to learn and understand everything about the world around him was not confined to engines, and he shared his father's love of nature, especially of birds. He encouraged robins to come to his engine to feed on the crumbs from his lunch, the way his father always did.

George Stephenson was eighteen years old, with a responsible job running his own engine and earning good wages. This would have been enough for most men – his father had settled for less – but it was not enough for him. Like many other young men of his years he probably had no exact idea of what he wanted to do with his life, but knew there was something better. Painfully aware of his own ignorance through lack of any schooling, he desperately wanted to better himself. The first step he took was to learn to read, so he could unlock the universe of knowledge hidden away from him in books. He also realised he must be able to write, and to understand arithmetic. To this end he began attending evening classes in the village of Walbottle, given by a local schoolmaster and attended by a few of the other colliery workers. Stephenson attended three classes a week, which could not have been easy after working a twelve-hour shift, but he persevered with a dogged determination to succeed. He used every spare moment to practise his lessons, scribbling on a slate during quiet periods with his engine. In later years, when addressing working men's groups as he so often did, he always advised the young men in his audience to 'Do as I have done – persevere.'

Although running an engine carried more responsibility than being a fireman, the best-paying and most responsible engine job at a colliery was that of brakesman of the pithead engine. This was used for hauling large baskets of coal, called corves (singular, 'corf'), to the surface, using a thick rope wound on a drum. The same engine was also used for lowering men, and equipment, to the bottom of the shaft, and for returning miners at the end of their shift. The brakesman had to halt the engine at the precise moment the load came level with the surface, and this needed the quick responses of a steady and reliable man. It was a difficult skill to master, but Stephenson was fortunate in being able to practise on a small winding engine operated by one of his friends. Eventually he got the position of brakesman at the Dolly Pit mine, near the village of Black Callerton, some two miles from Walbottle.

The new job gave him a good deal of free time between hauls, especially during the night shifts, and he put this to good use. In addition to working on his lessons he repaired shoes, a part-time occupation he had been practising for some time, to earn a little extra money. From this he graduated to shoemaking, and became quite expert. He was earning about £1 (£40) a week as brakesman, which was a good wage for a young man of twenty, and the shoe business allowed him to put more money aside. The reason for his thrift was that he had begun courting. The young lady, Fanny Henderson, was twelve years his senior.

Stephenson did not go drinking with the other men and was never seen drunk, which could not be said of many of his workmates. Nevertheless, he was popular, and was a particular favourite among the younger boys, probably because he treated them fairly and kindly. Some of the other workers were less considerate, and one in particular, an aggressive bully named Ned Nelson, used to terrorise the colliery. On returning to the surface at the end of one shift, Nelson swore at Stephenson for his alleged clumsiness with the brake. Stephenson would not be intimidated by anyone, and when Nelson threatened to kick him, Stephenson told him to go ahead and try. Nelson could not allow such defiance to go unanswered and challenged him to a public fight. A time was settled upon, several days hence.

News of the fight spread like pit fire, and although almost everyone was on Stephenson's side, most of his workmates felt too intimidated by Nelson to be openly supportive. A steady trickle of visitors came to the engine house to verify the truth of the story, and to wish Stephenson well, though the general opinion was that Nelson, who had a fearsome reputation as a pugilist, would slaughter him. 'Aye; never fear for me,' Stephenson assured them. 'I'll fight him.'

Nelson took time off work on the day of the fight, to make sure he was well rested, but Stephenson, seemingly unconcerned by the whole thing, continued working as if nothing untoward was about to happen. Finishing his shift at the end of the day, the same as usual, he made his way to the appointed place in the field beside Dolly Pit. Nelson was already there, baying for blood and swaggering before the expectant crowd in blissful contemplation of beating his opponent senseless. He planned to make it a punishing humiliation, as a lesson to anyone who dared defy him.

Bare-knuckled brawling was second nature to Nelson, but it was Stephenson's first-ever fight. Stripped to the waist like his adversary, he looked an unpromising match with his wiry frame and benign demeanour. But if Stephenson was feeling apprehensive he did not show it, and when the fight began he set about the unsavoury business as if he knew what he was doing. Within a few rounds it seemed he knew exactly how to conduct himself because Nelson was getting the worst of it. Stephenson carried on pounding away relentlessly, and went on to inflict a resounding defeat upon his astounded opponent. Shaking hands at the end of the fight, they became good friends.

When Stephenson left Dolly Pit for a new job as brakesman, he had saved enough money to furnish the top floor of a small cottage, at Willington Quay, for his bride-to-be. They were married late in November 1802. The twenty-one-year-old signed his name in the register with the deliberation of someone unused to writing, and the signature of Frances Henderson, penned beneath his own, appears to have been written by the same unsure hand.

Fanny must have thought she had married a very studious young man because he spent most of his evenings studying mechanics and building models to test his various ideas. He was especially interested in understanding the principles governing atmospheric engines, and in learning about James Watt's innovative modification. He was also intrigued by the idea of perpetual motion – like so many others of his time, he was beguiled by the notion that a machine could be built that, once set in motion, would continue running for ever, without consuming any energy. The idea of perpetual motion, like Ericsson's quest for a flame engine, was a delusion, rooted in a fundamental misunderstanding of how energy is exchanged in machines. He spent time on other futile quests too, and in later years lamented the loss of so much time, all of which could have been avoided had he been more knowledgeable. But Stephenson was no idealistic dreamer, and devoted much of his spare time to earning extra money from his shoe business. He also added clock and watch repairs to his services. This arose from a fortuitous chimney fire when the soot that filled their cottage clogged their prized wall clock. Dismantling the clock to clean the moving parts not only made it work again, but also familiarised him with the clockwork mechanism, encouraging him to tackle other repairs.

Fanny delivered his one and only son, Robert, on 16 October 1803. George was a proud and affectionate father, doting on his son as he did upon his wife. They moved to Killingworth the following year where he took up the post of brakesman at West Moor Colliery, one of the biggest and richest mines in the area. The move seemed to improve Fanny's health, which had been in serious decline for some time, but this was only temporary, probably attributable to her having become pregnant. Their second child, a daughter, was born on 13 July 1805, but she died three weeks later. Her father's grief was made all the more unbearable by the sight of his wife, with her dark sunken eyes and pallid frail body, racked by a terrible cough. Fanny died of consumption the following spring. Stephenson was devastated, and sought solace in hard work.

Shortly after Fanny's death he received an invitation to superintend a Boulton & Watt atmospheric engine in Montrose, Scotland. Now that Watt's original patent had expired increasing numbers were being built, so this would be useful experience for the future. The wages were good too, so he decided to take the job, and arranged for Robert to be looked after by a trusted neighbour. Packing a few possessions into a knapsack, he set off for Montrose on foot, the 200-mile journey probably serving as a catharsis for his grief.

Stephenson stayed in Scotland for only about a year, returning to his old job at Killingworth in 1805, the same year as Nelson's decisive naval victory at the battle of Trafalgar. During his time away he managed to save £28 (£1,500). His homecoming should have been a joyous occasion, but it was marred by the news that his father had been seriously injured at the colliery and was no longer able to work. His father had been making repairs inside an engine cylinder when a fellow worker, not realising anyone was inside the engine, opened the steam valve. The blast of steam struck him full in the face, permanently blinding him and causing extensive burns. If the colliery had paid him any compensation it was only a pittance because he was already some £15 (about £700) in debt. Paying off the debt from his Montrose savings, Stephenson moved his parents into a cottage and supported them for the rest of their days.

The period around 1807–08 seems to have been a particularly low point in George Stephenson's life, due in large part to the worsening conditions on the home front. The continuing war with Napoleon was a serious financial drain on the nation's resources, and William Pitt's government responded by imposing heavy taxes on consumables and introducing a graduated income tax. Prices, already high, kept rising, and wages, far from keeping pace, fell further behind. There was no recourse to collective bargaining because trade unions had been outlawed by the passage of the Combination Acts of 1799 and 1800, enacted in the government's belief that unions were the most likely breeding grounds for social unrest.

There was spasmodic rioting in towns throughout the land as desperate people demonstrated against high food prices and low wages. Lord Castlereagh, Pitt's Minister of War, responded by establishing a local militia of 200,000 men to maintain law and order at home. Meanwhile men were still being pressed into service to fight Napoleon overseas. Stephenson was eventually called up for military service. According to the law, he could either join the army or find a substitute to serve in his place. Choosing the lesser of two evils, he found a military man who was willing to be his substitute, on payment of 'a considerable sum of money'. This took most of his remaining savings. He was desperate enough to emigrate to America with his sister Ann and her husband, but he did not have enough money for the fare.

The change in his fortunes began so subtly that it would have completely escaped his attention. As he recounted many years later, 'I was trusted in some small matters, and succeeded in giving satisfaction,' which led to his being given increasingly larger responsibilities. The first step was taken when Stephenson noticed that the rope of his winding engine was wearing out in about a month, compared with about three months for similar engines in other pits. The price of new rope had escalated along with everything else, making replacement an expensive proposition. He attributed the premature wear to a misalignment of the pulleys, and when he suggested a remedy the head engineer encouraged him to go ahead and make the necessary changes. The marked improvement in the longevity of the rope was duly noted. But his next project was such a major contribution to the colliery that his employers were left in no doubt of his outstanding worth.

In 1810 the Grand Allies sank a new mine near Killingworth, called High Pit, installing a Newcomen atmospheric engine to pump out the water. Stephenson watched the installation with great interest, returning periodically to see how the new engine was running. But for reasons nobody could understand, not even the most experienced engineers, the engine's performance was so poor that the water level hardly dropped, and after almost a year of continuous pumping the water level was still too high for miners to enter the pit. The Grand Allies were justifiably displeased. Not only had they wasted considerable funds on an engine that did not work properly, but their investment in the new pit had not yielded a single knob of coal.

Stephenson, thinking he knew the cause of the problem, paid another visit one Saturday afternoon after work, for a closer look.

'Weel, George, what do you mak' o' her?' asked Kit Heppel, one of the miners. 'Do you think you could do anything to improve her?'

'Man, I could alter her and make her draw,' George replied thoughtfully, 'in a week's time from this I could send you to the bottom.'

Later on that afternoon Heppel mentioned his conversation to Ralph Dodds, the head viewer, who was at his wits' end. Knowing of Stephenson's reputation with machinery, he called at his home that same night.

'We are clean drowned out . . .' said Dodds desperately. 'The engineers hereabouts are all bet [beaten] . . .'. When Stephenson said he thought he could put the engine to rights, Dodds asked him to start work right away, assuring him that 'if you really succeed in accomplishing what they can not do . . . I will make you a man for life.'

Stephenson did not have the same reservations about working on the Sabbath as Hackworth, and began tackling the problem early the next morning. Word travelled fast, and when the engineers found out that a mere brakesman was going to try and resolve the problem that had eluded them, they became quite resentful. This could have been a problem for Stephenson, but he had anticipated their reaction and had reached a prior agreement with Dodds that he would choose his own men to assist him. This did not sit well with his 'betters', but Dodds ordered them to stand aside, and, with much grumbling, Stephenson and his crew were allowed to get on with the job.

Atmospheric engines are temperamental creatures that can be exhilarating, then completely exasperating. Affected by the weather, they work better on bright high-pressure days than on low-pressure wet ones. They can be working perfectly on some occasions, then refuse to start on others, and for no apparent reason. Their operators have to nurse, coax and cajole them, interpreting every little wheeze and clank and groan. George Stephenson knew his engines well, and it did not take him long to find that the main reason for the engine's poor performance was the inadequacy of the water spray used to condense the steam prior to the power stroke. Stephenson resolved this by enlarging the injection cock to almost twice its original diameter. He also raised the reservoir of the condensing water by 10 ft, thereby increasing its pressure.

He found the boiler would withstand a higher pressure than its normal 5 psi, so, contrary to all recommendations, he planned to double the pressure. As it was an atmospheric engine the raised pressure would not affect the force on the piston during the down stroke, when the condensing steam formed a partial vacuum inside the cylinder. However, when the higher-pressure steam was injected into the cylinder it would exert a significant force upon the piston during its up stroke. For example, if the cylinder were 4 ft in diameter, an average size for an atmospheric engine, a steam pressure of 10 psi would exert a force of about eight tons on the piston. This would have more than compensated for the weight of the piston and the inertia of the beam, making the up stroke active rather than passive, so causing the engine to run faster. Stephenson made a few other adjustments, and would have methodically cleaned all the working parts before reassembly – he was very thorough.

The job took four days, and by the following Wednesday steam was raised and the engine was all ready to go. A small crowd of expectant onlookers gathered beside the huge engine, most of them hoping to see it come to life again. Perhaps a few of the older hands were hoping for the worst, just to put the young upstart in his place. Stephenson was probably feeling confident the engine would perform well because he had methodically assessed the problem and effected a solution, but there must have been some anxiety when the cock was opened and steam hissed into the cold cylinder.

Stephenson's detractors must have been delighted when the engine took its first few strokes, because it began moving so violently that the entire engine room began to shake.

'Why, she was better as she was,' exclaimed a distraught Dodds, 'now, she will knock the house down.'

Stephenson probably realised what the problem was and remained calm. Because the pump was still empty of water it offered little resistance to the engine, which therefore began racing. But as the pump filled and began drawing water up from the flooded pit, the engine settled into a more sedate rhythm.

By ten o'clock that night the water in the pit was already lower than it had ever been. The engine kept on pumping all the next day, and by the Friday afternoon the pit was sufficiently dry to allow the pitmen to be sent down. Dodds was so delighted that he gave Stephenson a bonus of £10 (£420). It was a trivial sum compared with the great saving to the colliery, but Stephenson was enormously proud of the payment because it was the first recognition of his skill as an engineer. It was also the largest sum of money he had ever earned at one time. But his biggest reward was that Dodds promoted him to the post of engineman at High Pit, and increased his wages.

Word of his success quickly spread, and he was soon attending ailing engines throughout the Killingworth area. The more machines he dismantled and put to rights the more his knowledge grew. He continued reading mechanics during his spare time, along with repairing clocks and shoes, and building all manner of contrivances. Among these was a mechanised scarecrow for his vegetable garden that moved its arms in the wind. Stephenson had become a keen gardener and took great pride in growing the biggest cabbages and leeks in the village. They always won first prize at the local fête – winning was very important to George Stephenson.

Industrial accidents were not uncommon in Stephenson's day, especially in collieries, and in 1812, the year after his appointment as engineman, Old Cree, the engine-wright at Killingworth, was killed on the job. Unfortunate as this was for Cree and his family, it was fortunate for Stephenson because he got his job, at a salary of £100 (£3,800). Aside from visiting the collieries operated by the Grand Allies to inspect the machinery, he began erecting engines himself, starting with a winding engine at High Pit. But he did not confine himself to engines, and one of his most innovative ideas was to install a self-acting incline for running wagons of coal down a hill to a loading area. The wagons were hitched to a thick rope looped around a drum, and as the loaded wagons descended the incline, their inertia was used to haul the empty wagons up to the top. While this was not the first time the system had been used, it was one of the first self-acting inclines operating in the Killingworth district. Another of his ideas was to install a winding engine underground in the Killingworth Pit, for hauling coal wagons up an incline. This changed the entire working operation of the pit, reducing the number of horses needed for hauling coal from 100 to about fifteen, with a considerable reduction in expenditure. Stephenson's innovations touched a sentimental chord with the Grand Allies – money – and Sir Thomas Liddell, whom he often encountered while making his rounds, warmly encouraged his efforts.

Although Stephenson was earning comparatively good wages, he continued with his spare-time jobs and managed to accumulate £100 (almost £4,000). His main reason for saving so hard was to provide a good education for Robert: he was determined his son would never have to endure the disadvantages he had suffered through lack of schooling. Robert had been attending the village school, but he soon outgrew its rudimentary teaching, and when he was about twelve his father enrolled him in a school at Newcastle. This was a distance of six or seven miles, and Stephenson bought him a donkey for the journey. Every morning young Robert set off for school, dressed in the grey suit his father had made for him, his school bag slung over his shoulder. Stephenson took the opportunity of learning along with his son, and they would often go over the day's lessons at night.

One of the practical things they did together was to build a sundial during the school holidays, which they mounted over the front door of their Killingworth cottage.

George Stephenson did not suffer Trevithick's compulsion to dash from one project to the next, but he did share his interest in finding new solutions to old problems. One that had been occupying his attention was the haulage of coal from the colliery to the shipping wharf. This involved the use of hundreds of horses, and if they could be eliminated the savings would be enormous. Installing a series of winding engines along the wagonway was one solution, but his visits to Wylam and Coxlodge had convinced him that the way of the future lay in the new travelling engines. Stephenson voiced this opinion to Sir Thomas Liddell, convincing him to authorise the construction of a locomotive. 'Lord Ravensworth and partners were the first to entrust me with money to make a locomotive engine,' Stephenson recounted many years later, and 'That engine was . . . called . . . "My Lord".'

The usual assumption is that the locomotive was named after Liddell, but he did not become Lord Ravensworth until 1821. However, one of the other Grand Allies, the Earl of Strathmore, was a peer of the realm, so *My Lord* may have been named for him. Work on *My Lord* probably began some time late in 1813, in the colliery workshop at West Moor. The facilities were poorly equipped and far from ideal, but Stephenson had to work with what he had. The colliery blacksmith, John Thirlwall, was the leading mechanic on the project, and although he was a first-rate workman, like Stephenson, he had no experience in building locomotives. Both had to learn as they went along, making do with the few hand tools at their disposal, and improvising others when needed. There were some desperate moments, with as many frayed nerves as barked knuckles, but they got the job done, in about ten months.

The wrought-iron boiler was 8 ft long and just under 3 ft in diameter, with a simple straight-through flue. Stephenson probably avoided the Wylam preference for a return flue because of the technical difficulties. Two vertical cylinders, each 8 in. in diameter, were sunk half-way into the top of the boiler, and each piston rod was attached to a horizontal cross-head, lying parallel to the ends of the boiler. The up and down movement of each cross-head was transmitted to the wheels by a connecting rod and crank, through a series of cogged wheels, very much like the systems used in the Blenkinsop and Wylam engines. The four wheels were therefore permanently linked in a four-wheel drive. The tall chimney was a continuation of the flue, bent at right angles, and steam vented directly into the atmosphere rather than into the chimney. *My Lord* was an amalgam of the locomotives Stephenson had seen operating at Wylam and at Coxlodge, the main difference being that he used edge rails rather than plate rails or a rack-and-pinion system.

The locomotive was placed on the track on 25 July 1814 and, as the fire was lit and steam raised, Stephenson might have thought back to the recalcitrant engine at High Pit. But this time the engine was entirely new and untried, and built to his own specification, so if it failed to work properly there was no one to blame but himself.

Some doubts had been expressed whether the wheels would have as much traction on edge rails as they would on plate rails. But Stephenson was confident this would not be a problem because he had already conducted a simple traction test with a loaded carriage: several workmen had stood on the spokes of the wheels on one side and the wagon had inched forward, without any slippage.

When the working pressure was reached the regulator valve was opened to admit steam into the cylinders. *My Lord* moved forward, albeit hesitantly. Her movements were anything but smooth, and each piston seemed to be competing with its neighbour as it jerked its own pair of wheels forward. The lack of springs and the unevenness of the rails accentuated the jerkiness of its movements, giving her driver a bumpy ride. But she did make progress, moving at a brisk walking pace. Because of the difficulty of machining with the tools then available, engineering tolerances were poor and the cogged wheels did not mesh as well they might. As a consequence there was a good deal of mechanical noise from the transmission, but it was nothing compared with the blast of the venting exhaust steam. When she began making regular runs along the track there were many complaints from the local landowners, and one of them threatened legal action if the noise problem was not put to rights.

The line was on a gradient of 1 in 450, but the engine had no hesitation in hauling a load of 30 tons up the incline, at about 4 mph. Its steaming capacity limited it to moving no faster than a horse could walk, and its running costs were on an equine equivalent too. *My Lord* was not a dazzling success, but it was an excellent start, especially for a man with no education, formal training or experience in building locomotives. And Stephenson continued to enjoy the confidence of Sir Thomas Liddell, a man whose pockets were deep.

Within a few months of completing his first locomotive Stephenson was working on improvements for a second. The transmission of power through the intermeshing cogs had to be simplified, and Stephenson, working with Dodds, devised a system whereby a long rod from each piston's

This locomotive, built by Stephenson for the Hetton Colliery, was similar to his second engine.

cross-head linked directly with its corresponding wheel. But they had to overcome the problem of the unevenness of the rails, otherwise the linkage to the wheels would be dislocated when the first major bump was encountered. This was resolved by fitting ball-and-socket joints at both ends of the long connecting rod. Dispensing with all the gearing meant that the two pairs of wheels were independent, but they wanted to keep them linked, to maintain four-wheel drive. They planned to do so by installing cranked axles, linking them with a rod, but 'the mechanical skill of the country was not equal to the task of forging cranked axles to the soundness and strength necessary . . .', so they had to use a chain instead. Stephenson and Dodds patented their improvements on 28 February 1815, describing the cranked axle as one of the methods of linking the front and back pair of wheels. This was the first time a cranked axle was proposed for a locomotive, but it would be more than a decade before it became a reality, in *Novelty* at Rainhill.

The second engine was completed in 1815, the year Napoleon was defeated at the battle of Waterloo. In addition to the improvements included in the patent, the exhaust steam was discharged into the chimney. Aside from reducing the noise this increased the draught to the furnace, and 'the power of the engine became more than doubled . . .'. It is unclear whether this second engine was named *Blücher*, after the Prussian general who saved the day at Waterloo, or whether *My Lord* was renamed *Blücher*, but the second locomotive was a success and several more were built, some of which continued operating at Killingworth for more than forty years.

Stephenson realised that the future development of the locomotive was severely hampered by the weakness of the rails, and that significant improvements had to be made. At that time only cast-iron rails were available, and these were produced in only 3 ft or 4 ft lengths, making for numerous joins in the track. Engineers were already devising ways of testing metals, and knew that cast iron, which was more brittle than wrought iron, had only about half the tensile strength – the resistance to being pulled apart. Rail breakages were quite common, but it had been found that the strength of a length of edge rail could be increased by making it deeper in the middle, diminishing in depth towards either end. The resulting 'fishbelly' rails first appeared during the early 1800s, though it took engineers another decade or so to comprehend the underlying theory. But there was still much room for improvement, and to this end Stephenson was seconded to the Walker Ironworks, in Newcastle, to work two days a week on rail improvements with William Losh, the senior partner. For this Stephenson was paid an additional £100 (£4,750) per annum, indicating the high regard in which he was held by his employers.

The effects of the unevenness of the rails would be greatly reduced if the wheels were sprung, so they could track the variations in rail level. Remarkably, though, steel springs of the required strength were unavailable. Stephenson's innovative solution was to set four cylinders into the underside of the boiler, one for each wheel, attaching each piston rod to its respective wheel axle. The force generated by the four pistons was not enough to support the weight of the locomotive, and the axles were still attached to the boiler frame. But the pistons acted as springs, keeping the wheels pressed against the rails despite the unevenness.

Regardless of how well the 'steam springs' may have functioned in practice, the brilliance of the concept, which Stephenson patented jointly with Losh, leaves no doubt of his ingenuity. He was much more than a practical engineer who had worked his way up from the obscurity of the mines – he was an original thinker.

This is clearly shown by his invention of a miners' safety lamp in 1815. Space does not permit detailing the remarkable story, but after experiencing several devastating explosions, caused by the ignition of leaking methane gas, he devised a lamp that could be safely used underground. With total disregard for his own safety, he tested the prototype in one of the most dangerous galleries in the pit, where methane hissed loudly from the roof.

His Geordie lamp, as it was called, pre-dated the better known Davy lamp by weeks, but Sir Humphrey Davy, who received a premium of £2,000 (£90,000) for his work, arrogantly denied Stephenson's claim to originality. An outraged Earl of Strathmore, one of the Grand Allies, gave Davy short shrift for his condescension, and a subscription list was launched that raised £1,000 (£45,000) for the local hero. Stephenson was presented with the award early in 1818, at a dinner held in his honour at the assembly rooms in Newcastle. Knowing he was expected to give a speech, he wrote out the text of what he wanted to say. Replete with spelling and grammatical errors, it began:

> Sir – In Receiving this valuable present which you and the Gentleman of this Meeting has bean pleas'd to present me with this day I except with Gratitude. But permit me to say valuable as this present is and gratefull as I feal for it I still feal more by being honour'd by such and highley respectable meeting the Gentlemen of which having not only rewarded me beyond any hopes of mine for my endeavours in construting a safity Lamp but supported me in my claims as to priority in my invention to that of that distinguished Pholosipher S H Davy . . .

At a family gathering some time after the presentation, he confessed to being so embarrassed during his speech that a candle could have been lit from his face. When one of his sisters-in-law, joking about his elevated station in life, said he would now start wearing shirt ruffles and looking down on his friends, he solemnly replied that he would never change.

On 19 April 1821, George Stephenson accompanied by his friend Nicholas Wood, travelled to Darlington to see Edward

Pease, a meeting that would prove pivotal in both Stephenson's career and the history of steam locomotion. Stephenson intended to offer his services in the construction of the Stockton & Darlington Railway, and Wood, who knew Pease, joined his friend to make the introductions. According to the more imaginative accounts of the historic meeting, the two Killingworth rustics had called upon Pease on the off-chance of being seen, and Wood, the more articulate of the two, was there to speak on behalf of his brilliant but bumbling friend. Stephenson may have been bashful at the assembly rooms dinner, but a great deal had changed in the intervening three years. Stephenson was now widely known and respected in the north of England as an engineer in his own right. His £100 salary from the Grand Allies had long since become a retainer, and he continued receiving £100 from Losh's company. He also received a salary from the Hetton Colliery Company, for whom he had built a railway for hauling coal. Stephenson had also been wise in the investment of the safety lamp premium: his assets included part ownership of a colliery. His financial security and growing reputation as an engineer had done much for his self-confidence, and he would have been quite at ease in the company of an influential man like Pease.

Edward Pease, fourteen years his senior, was cut from the same down-to-earth homespun as Stephenson, and the two men got along well. According to Pease's recollection of their first meeting, 'There was such an honest sensible look about George Stephenson, and he seemed so modest and unpretending, and he spoke in the strong Northumberland dialect.'

Pease, a shrewd and wealthy businessman, had diversified beyond the family's interests in the woollen trade into transport. Four years earlier his home town, and the

Hetton Colliery, sometime during the early 1820s, showing early Stephenson locomotives hauling coal.

neighbouring town of Stockton, had begun serious discussions on a transport link to carry coal from the Darlington mines to the shipping wharves at Stockton. Roads were improving in Britain, largely through the adoption of John MacAdam's new methods of construction. The turnpikes were the best, and by the 1820s there were about 20,000 miles of these toll roads throughout the country. Although much faster than the canals, they could not carry heavy freight and so were not an option for coal haulage. Pease, long convinced 'that a horse on an iron road would draw ten tons for one ton on a common road', became an advocate of a wagonway. Most of the Stockton residents preferred a canal, but Pease's arguments prevailed, and he became a founding director of the Stockton & Darlington Railway Company.

During their meeting Stephenson told Pease of the success of locomotives, and how they would be far more satisfactory than horses for the Stockton & Darlington Railway. Pease was a progressive man, and having listened with great interest, he arranged for a deputation to visit Killingworth. Days later the deputation reported back in favour of locomotives. Pease accordingly invited Stephenson to plan the line, asking him what remuneration he expected. Stephenson valued his services at £300 (£18,000) per annum, together with expenses.

Among the other visitors to Killingworth that year was William James, a lawyer who acted as land agent for large estates. James, a staunch advocate of the locomotive, could visualise a day when the entire country would be united by railways, a notion shared by Stephenson. James was so inspired by his first Killingworth visit that he made the prophetic statement 'Here is an engine that will . . . effect a complete revolution in society.' After meeting Stephenson on a subsequent visit he declared he 'was the greatest practical genius of the age . . .'. James expressed the same sentiment in a letter to Pease, telling him that Stephenson's locomotives were superior to all the others. Pease was probably well pleased by such a reassurance. Earlier that summer James had met Joseph Sanders, an influential Liverpool merchant. Exasperated by the poor service offered by the canal linking Liverpool with Manchester, Sanders was campaigning for the construction of a railway. He was so impressed with James's knowledge of the subject that he commissioned him to undertake a survey of the proposed line.

Pease invited Stephenson to conduct a resurvey of the Stockton & Darlingon line, and he was officially appointed as surveyor on 28 July 1821. This would be only his second survey, the first having been made two years earlier for the Hetton Railway, linking their colliery with the shipping wharf.

A railway survey was more than a detailed map because it had to include a vertical profile through the entire route, showing all gradients and obstacles to be negotiated. This was important because locomotives, unlike modern road vehicles, cannot climb steep gradients. Stephenson had to learn not only how to use a surveyor's tools and techniques, but also how to choose the best routes, negotiate obstacles and construct inclined planes. And he acquired all these skills without any formal education, while holding down a full-time job at a colliery.

Just a few days before Stephenson began the survey Pease visited Killingworth, to see the locomotive for himself. He was much impressed, as he explained in a letter to his cousin, the influential London banker Thomas Richardson. The letter is dated 10 October 1821:

> the more we see of Stephenson, the more we are pleased with him . . . he is altogether a self taught genius . . . there is such a scale of sound ability without anything assuming . . .

Pease was also quick to perceive the enormous potential of steam locomotion:

> don't be surprised if I should tell thee, there seems to us . . . no difficulty of laying a rail road from London & to Edinburgh on which waggons would travel & take the mail at the rate of 20 miles per hour . . . we went along a road upon one of these Engines conveying about 50 tons at the rate of 7 or 8 miles per hour, & if the same power had been applied to speed which was applied to drag the waggons we should have gone 50 mile per hour . . .

He finished his letter with the prophetic statement: 'we shall make a good thing of this concern'. Stephenson was a diamond in the rough, but Pease, with Richardson's assistance, would groom him for the role he was to play. A little over three years after his Killingworth visit Pease wrote of Stephenson that:

> he is a clever man, but he must have leading straight, he should always be a gentleman in his dress, his clothes real and new, and of the best quality, all his personal linen *clean every day* his hat and upper coat conspicuously good, *without dandyism* . . .

Stephenson began surveying the line in the autumn, assisted by John Dixon, appointed by the Stockton & Darlington Railway Company, and by his eighteen-year-old son Robert. Stephenson was a hard taskmaster, starting at daybreak and working on until it was too dark to continue. Dixon recalled one particular occasion, after a long day's work, when the surveying party were about to go their separate ways for the night. Stephenson had arranged to sleep at a local farmhouse, while some of the others were going to walk four miles into Darlington. Stephenson reminded them to set off before daybreak the following day, so they could start work directly it was light.

Stephenson was quite unconcerned for his own comfort during the survey and made no plans for meals, taking his chances at getting a drink of milk or a hunk of bread at a cottage along the way. He was occasionally invited to join a family meal around the kitchen table and his friendly face became quite a familiar sight in the neighbourhood, where he was always received with a warm welcome.

The work went well and was finished in less than three

weeks. Soon after submitting his completed survey he was appointed engineer, at an annual salary of £660 (£45,000). Being the first engineer of the first public railway in the world was a remarkable achievement for a man who had spent the first quarter-century of his working life in a colliery.

Stephenson, not good at delegating, took responsibility for everything. He negotiated contracts with local workmen to cut the embankments, located quarries to supply stone for the rail blocks, designed the inclined planes, determined what sort of rails to use, and sundry other things above and beyond his primary role of overseeing the entire project. At that time he had a business arrangement with Losh for manufacturing cast-iron rails, and stood to make a good deal of money in supplying these to the railway. Then he discovered that Michael Longridge's Bedlington Ironworks was producing wrought-iron rails in 15 ft lengths. Although more expensive than cast-iron rails they were considerably tougher, and required fewer joins because of their longer length. Without any consideration for his own financial interest he recommended using Longridge's rails. Losh was neither understanding nor forgiving, and it marked the end of their business relationship.

Stephenson remarried, in the spring of 1820. Elizabeth Hindmarsh was said to have been a former sweetheart, whose father had refused to let her marry because of Stephenson's lowly position. But her brother denied this as idle gossip.

Writing to a friend a few months before his wedding, Stephenson recalled how he had been doing some soldiering. 'We have about 3 hours drill every day & then plenty to eat & drink at Gosforth house.' The call to arms was in response to the civil unrest at home – since the restoration of peace following Napoleon's defeat at Waterloo there had been renewed agitation among reformers for an improvement in the lot of the working classes. George Stephenson, well acquainted with the hardships and injustices suffered by his own class, might have been expected to sympathise with the reformers. However, not all working people were caught up in the reform movement, and he was not turning his back on his working-class roots by siding with law and order. Besides, he was joining ranks with the industrialists, and knew which side his bread was buttered.

Four thousand miles away the frigate *San Martin* sank at anchor in Chorillos Bay, off the Peruvian coast. The loss of a tired old two-decker was probably no great financial loss, but she did carry a large number of brass cannon, and there was a valuable cargo of copper and tin in her hold. Richard Trevithick, now fifty-one, negotiated a salvage contract with the government in Lima: in exchange for recovering the brass cannon, he would keep all the cargo. Using a 'rude diving bell', he recovered about £2,500 (£170,000) worth of copper and tin. One of his English friends urged him to send at least £2,000 home to his wife, but, instead, 'he embarked the money in some Utopian scheme for pearl fishing at Panama, and lost all!'

The five years Trevithick had spent in Peru had been neither dull nor predictable. During the first year, two of his men were robbed by their guides as they slept, then murdered by having their skulls crushed with boulders. Uville had turned out to be almost as treacherous as the guides, and the man who had been so courteous and respectful in England was resentful and jealous of Trevithick from the moment he arrived. Claiming sole credit for introducing steam engines to Peru, he tried to take control of the entire mining operation, opposing Trevithick at every opportunity, and undermining his position. The Cornishman soon lost patience and struck off on his own to make his fortune. But he had only been gone a year when he received news of Uville's death, and an entreaty to return. Trevithick agreed, and within a short time of his return the mines at Cerro de Pasco were 'in the most prosperous state . . .'. Things went well for a time, but Trevithick eventually lost patience with the petty jealousies and in-fighting, and not even an offer of $8,000 per annum could persuade him to stay.

His first enterprise was with a group of miners who were extracting silver from its ore in the traditional way, using mercury. After bonding with the silver to form an amalgam – the same as dental filling – this was heated in evaporating rooms, driving off the mercury and leaving the silver. The toxic mercury vapour killed the workers within a year or so and Trevithick wanted them to use the British method of grinding and smelting the ore. Investing some of his own money in the machinery, he hoped to see a good return, but, like most of his schemes, this one failed, and he lost his entire investment.

Peru was still under Spanish control, governed by a viceroy from Spain. Soon after Trevithick's arrival the viceroy granted him a special passport so that he could travel the country advising miners. In return for his services the government granted him possession of any unclaimed ore deposits. Trevithick settled upon a large and easily accessible 'copper vein' which he believed could be 'obtained without risk of capital'.

At about the time Trevithick began mining the copper, Simon Bolivar, the liberator of Venezuela, rode into neighbouring Ecuador, to the north, driving out the Spanish and pushing south. 'When the patriots arrived in Peru,' Trevithick wrote, 'the mine was deserted by all the labourers, in order to avoid being forced into the [Spanish] army.'

Work came to a halt, but when the Spaniards retreated into the interior and the threat of impressment passed, his work force returned and mining resumed. When the Spaniards returned his workers fled again, but by this time Trevithick had stockpiled about £5,000 (£340,000) worth of copper ore, with plenty more still in the ground. 'However, revolution followed revolution, and the war appeared to me to be interminable. Even Bolivar's arrival at Lima made it still worse, for he forced me into the army . . .'. Bolivar's cavalry was short of firearms, so Trevithick designed a carbine, a short-barrelled firearm suitable for firing from horseback. He appears to have got on well with Bolivar, and would visit him at his residence. During one of his visits he learned of an assassination plot against Lord Cochrane, a fellow Englishman with whom he was acquainted.

Lord Thomas Cochrane (1775–1860), later the tenth Earl of Dundonald, was four years younger than Trevithick and even more hot-headed and independent. Born of an aristocratic but impecunious family, he had joined Nelson's navy, serving with distinction. His brilliant actions won him a knighthood from the king and adoration at home, where he successfully stood for a seat in Parliament. Always on the side of the underdog, he became a staunch reformer and rocked the Tory government by exposing the most flagrant abuses of patronage among its members. But he fell from grace in a stock exchange fraud, of which he was almost certainly innocent. Found guilty, he was fined, gaoled, stripped of his knighthood, and struck from the Navy List. Cochrane was so embittered that when South American patriots invited him to take command of their navy, he jumped at the opportunity.

Among his early successes was the capture of the Spanish flagship, the forty-four-gun frigate *Esmeralda*, in an action described as the most brilliant example of a cutting-out expedition in naval history. Cochrane was hailed a hero and liberator, but his dazzling successes fomented jealousies among the leadership and the assassination plot against him may have been instigated by any one of his enemies. In learning of this, Trevithick swam out to the *Esmeralda*, anchored near Lima, to warn him, probably saving his life. Cochrane eventually returned to England, where his creative mind turned to experimenting with steam locomotion.

Army life had little appeal for the free-spirited Trevithick, and he resolved to do something about it. An opportunity arose when Bolivar sent him to Bogotá on a special mission. Stopping *en route* at Guayaquil, on the Ecuadorian coast, Trevithick heard stories of the fabulous riches of the mines of Costa Rica. This was an opportunity too good to miss, so he deserted from the army and set off, once more, to make his fortune. His only regret in leaving Peru was that he had to abandon his copper mine, and all the ore he had collected, together with all his equipment.

The year after Trevithick left Peru, Stephenson entered into a partnership with his son, Robert, Edward Pease, Michael Longridge and Thomas Richardson, to manufacture locomotives and stationary steam engines. Officially named Robert Stephenson & Co., they were usually known as the Forth Street Works, after the Newcastle street close to the factory. And it was there that Robert Stephenson, George's sole descendant, built *Rocket*.

CHAPTER 6

Famous son of a famous father

Robert Stephenson's departure on a three-year contract as engineer-in-chief of the Colombian Mining Association, barely a year after the formation of Robert Stephenson & Co., must have struck many people as odd. It has been suggested that Robert wanted to leave England because of a rift with his father over the latter's poor treatment of William James. Robert had formed a close friendship with James while they worked together on a preliminary survey of the Liverpool & Manchester Railway. James was to have completed a detailed survey of the line, but, owing to his overextending himself, both financially and temporally, the company was obliged to dispense with his services: James, like Trevithick, was for ever taking on too many projects and exposing himself to financial risk. George Stephenson had little patience with such dalliance and readily agreed to take his place. Robert, empathetic to a fault, probably had some misgivings over his father's lack of sympathy. But it seems unlikely to have caused a rift between them, and Robert always seems to have been on the best of terms with his father. Indeed, before sailing from England he wrote a hurried letter to his stepmother explaining how he had arranged for three-fifths of his salary to be paid to his father. George Stephenson was not a man without means, but his son was well aware that the Forth Street Works were a heavy drain on his resources.

Robert Stephenson lost his strong accent, along with the vestiges of his parochial upbringing, in the ballrooms and drawing rooms of Bogotá and Mariquita, in Colombia. This was no passive transformation but a conscious and determined effort to cast off the social fetters he knew would always hold him back. His father would have wholeheartedly approved. Nobody was more aware than George Stephenson of the heavy burden of a northern burr, an accent few people in the south could understand: what was the sense of a proper education if folk knew *nowt* of what you were saying? Robert did have a semblance of proper education, as George once explained in a letter to a friend:

> I have onley [*sic*] one son who I have brought up in my profeshion [*sic*] he is now nearly 20 years of age. I have had him educated in the first Schools and is now at Colledge [*sic*] in Edinbro'. I have found a great want of education myself but fortune has made a mends for that want.

Robert spent only one term at Edinburgh University, where he studied natural history, natural philosophy and chemistry. The former, taught by Robert Jameson, was confined 'chiefly to Zoology, a part of Natural History which I cannot say I am enraptured with . . .'. Robert noted that natural historians spent 'a great deal of time in enquiring whether Adam was a black or white man', and he was unable to infer 'any ultimate benefit' of zoology, 'unless to satisfy the curiosity of man'. His introduction to natural philosophy seems to have been equally disappointing, and he did not bother writing out his notes after the lecture as they contained 'nothing but the simplest parts', with which he was already 'perfectly acquainted'. But he was 'highly delighted' with Professor Thomas Hope's lectures on chemistry, saying he was 'plain and familiar in all his elucidations'. Darwin, who similarly had a short studentship at Edinburgh just two years later, had a similarly high regard for Hope.

Regardless of his disdain for Jameson's lectures on zoology, he accompanied the professor on a geological excursion at the end of the term. In later years he recalled the experience with much pleasure. The freedom of hiking, knapsack on back, across the rugged Scottish countryside was reminiscent of the pleasures he had enjoyed the year before, when he had assisted his father in surveying the Stockton & Darlington Railway.

George Stephenson probably thought he had given his son a university education, but the few months Robert had spent at Edinburgh were little more than a finishing school for the three-year apprenticeship he had served with Nicholas Wood at Killingworth. Nonetheless, the upbringing and education his father had provided gave his son advantages he never enjoyed. George Stephenson taught his son the practical side of engineering, showing him how things worked and giving him an appreciation of the properties of materials. What he was probably unable to pass on was the ability to look at problems conceptually, and Robert may have acquired that through his formal education. Being able to stand back and ask questions gave Robert Stephenson an entirely different perspective on engineering problems from his father. George Stephenson saw things empirically whereas Robert looked at them analytically, and this would have enormous consequences on the development of the locomotive when he returned from South America.

Robert had worked hard during his apprenticeship with Wood, learning all aspects of coal mining, but his real vocation was with the embryonic railway, and he participated in a second survey of the Stockton & Darlington Railway just before going up to Edinburgh. Following his brief spell at university he assisted his father in his newly appointed role as engineer of that railway. This was shortly before the formation of the locomotive works that bore Robert's name.

Robert was very much his father's son, forthright and possessing the same down-to-earth simplicity. But while George used a forceful character to compensate for his social shortcomings, his son beguiled those he met with a genteel honesty that could be quite disarming. The expatriate gentlemen, with whom he became acquainted during his infrequent visits from the Columbian hinterland, must have been much taken aback when he asked them to correct his English. Every time he used an incorrect pronunciation or uttered a Northumbrian expression they were to tell him. He wanted them to be brutally frank in their admonitions, and was never offended by their most scathing criticisms.

The bamboo cottage where he resided lay in a rolling green paradise of tree ferns and magnolias, where humming birds darted among myriad flowers. This land of monkeys and macaws, of lianas and orchids, would have been an ideal location for a Darwin or a Banks to collect their specimens. But Stephenson was not a naturalist and had come to South America in search of mineral wealth.

Robert Stephenson's bamboo cottage, Santa Anna, Colombia.

The idea of going to South America was first seeded in Robert's mind by Thomas Richardson, one of the partners in the Forth Street Works and a close friend, as well as cousin, of Edward Pease. Like many others in the city, he was infatuated by the lure of South American gold and minerals. When he learned that the Colombian Mining Association was raising capital for a mining venture, he realised it would be an ideal opportunity for obtaining orders for the Forth Street Works. In addition to their requirements for steam engines, pumps and other mining equipment, the Association needed skilled workers. Richardson directed their enquiries for miners and other manual workers to George Stephenson, but when it came to an on-site engineer to oversee the project he immediately thought of Robert, whose abilities he held in high regard. The mere mention of joining an expedition to South America to search for gold and silver would have quickened the pulse of any twenty-year-old, but when the unattached young man was as resourceful and as bent on adventure as Robert Stephenson, the appeal was irresistible. If he needed a justification for going, the change in climate from the damp grey cold of Northumbrian winters to the luxuriant warmth of perpetual summer would work wonders for his constitution, which had never been robust. Since boyhood he had suffered a weakness of the chest that threatened to turn into the same consumptive disease that had taken away his mother. South America was too good an opportunity to miss, and he began preparations for the adventure ahead.

At the behest of the Colombian Mining Association he undertook a trip to Cornwall to study mining techniques, accompanied by his uncle Robert. The excursion was a valuable learning experience, but, in spite of Robert's youthful fervour for the Colombian adventure, his father remained unenthusiastic. Not even the annual salary of £500 (£29,000) could persuade him that the scheme had any merit. But Robert's letter to his father about his Cornish excursion may have helped tip the balance:

> when one is travelling about, something new generally presents itself, and . . . seldom fails to open a new channel of ideas . . . This I think is one of the chief benefits of leaving the fireside where the young imagination received its first impression.

In another letter to his father, Robert made a direct appeal for his approval:

> But now let me beg of you not to say anything against my going out to America, for I have already ordered so many instruments that it would make me look extremely foolish to call it off . . . You must recollect I will only be away for a time; and in the mean time you could manage with the assistance of Mr Longridge, who . . . would take the whole of the business part off your hands. And only consider what an opening it is for me as an entry into business; and I am informed by all who have been there that it is a very healthy country . . .

Inevitably, if reluctantly, his father gave his approval, telling Longridge that, 'the poor fellow is in good spirits about going abroad, and I must make the best of it'. Like Richardson, Longridge had the highest regard for Robert's abilities, and had become his close friend and confidant. Of all the business partners, Longridge was probably the most concerned over Robert's pending long absence from the locomotive works. In a letter to Robert, intended to be delivered before he sailed from Liverpool, he wrote:

> When your father and your other Partners consented that you should go to Colombia, it was with the understanding that your engagement was only of a temporary nature, and that as soon as you had informed yourself of the practicability of forming a Rail Way and had made your Geological inquiries you should return to England & make your report.
>
> *On no account* would we ever have consented that you should become the *Agent of* Messrs *Graham & Coy for 3 years.*
>
> I have spoken to our friend Mr Thomas Richardson upon the subject, and he promised me to use his

> influence . . . so that you may be released from your Agreement as soon as you have satisfied yourself upon the subject for which you are gone out and I am hopeful that you will be able to return to us in the course of the year 1825.

In spite of Longridge's sanguine hopes, Robert stayed in Colombia for the entire three years, not because he had been bitten by the same gold bug that had infected Trevithick or because of his love of South America, or of the job. In truth the job had been a vexatious futility from the very start. All the equipment that had been shipped out from England had arrived safely, but the heaviest pieces could be transported only by wagon, and since the precipitous mountain paths could be crossed only by mule, nothing but the smallest pieces could reach the mines. Stephenson informed London that other machinery should be shipped in pieces small enough to be carried by mule, but by the time his dispatch reached England the equipment was already at sea. On arrival in Colombia it was dumped on shore alongside the rusting remains of the previous shipment.

Making the most of a bad situation, Stephenson took the first contingent of miners, along with the few pieces of equipment they could carry, to the town of Santa Anna. After completing the treacherous twelve-mile journey they began opening the disused mines and working out the ore. But there was only so much that the small and ill-equipped gang could achieve, and it was several months before the second, and sizeable, body of miners arrived. They were all from Cornwall, but rather than improving the situation they made matters considerably worse. A more insolent and drunken rabble of men would have been hard to find. They had already outraged the morals of the local inhabitants, and had so incensed the Governor with their disreputable behaviour that he had lodged a formal complaint. 'I dread the management of them,' Stephenson wrote to the manager at Bogotá. 'They have already commenced to drink in the most outrageous manner.'

He hoped things would improve when they had settled in and begun to work, but it was not to be. One night a drunken mob occupied part of the building in which he was sleeping. And there they stayed, yelling threats and taunts at the 'beardless boy'. Their contempt for Stephenson was as much for his being from the north country as for his not having worked his way up from the bottom of a Cornish mine. Rising from his bed, he made his way through the drunken mob, stunning them into silence with his audacious action.

'It wouldn't do for us to fight tonight,' he told them, calmly, looking across the crowded room. 'It wouldn't be fair: for you are drunk and I am sober. We had better wait till to-morrow. So the best thing you can do is to break up this meeting, and go away quietly.' The mob maintained their silence for several long moments before finding their feet and shuffling outside. And there they remained for two or three hours, chanting their protests. But the sight of the engineer-in-chief through the open door, calmly smoking a cigar, defused the situation.

Things slowly improved over the next few months and he eventually won the respect of the men. But he could never get more than half a day's work out of them in any one day, and drunkenness always accounted for about a one-third absenteeism in the 160-strong work force. The situation was never going to change, and the letters he received from England during his first year confirmed that the futility of the Colombian venture was widely recognised by the more enlightened business minds in London. Faced with certain failure in the New World, and constant reminders from the Old that he was badly needed back at the locomotive works, he would have gladly returned home. But he stuck doggedly to the terms of his agreement: he had given his word.

Riotous assemblies were not the prerogative of South America, and in England's green and pleasant land Robert's father was faced with armed opposition to his survey of the Liverpool & Manchester Railway. In a postscript to a letter to Joseph Pease, Edward's son, who handled much of the business, George Stephenson wrote:

> We have sad work with Lord Derby, Lord Sefton, and Bradshaw the great Canal Proprietor, whose grounds we go through with the projected railway – Their Ground is blockaded on every side to prevent us getting on with the Survey – Bradshaw fires guns through his ground in the course of the night to prevent the Surveyors coming on in the dark – We are to have a grand field-day next week, the Liverpool Railway Company are determined to force a survey through if possible – Lord Sefton says he will have a hundred men against us . . .

Much of the resistance to the survey came from the Mersey & Irwell Navigation Company, which was protecting its vested interest in the Liverpool & Manchester Canal. Responding to criticisms made by the railway's proponents, they stated that the average travel time between the two cities was not thirty-six hours, as had been claimed, but only twelve, whereas it would take five or six hours by train. Shippers who used the canal knew that the twelve-hour claim was patently false. Furthermore, the railway was planning to complete the journey in a little over three hours. The canal company's rebuttals were informally transmitted to the Members of Parliament who were about to hear evidence for and against the passage of the Liverpool & Manchester Railroad Bill.

By the spring of 1825 the Bill had reached the committee stage and George Stephenson, as the surveyor and engineer of the proposed line, underwent the gruelling cross-examination mentioned earlier. Some of the fiercest opposition during his three-day ordeal came from the ranks of the most powerful landowners in the country, and they had staunch supporters in the House of Commons. Stephenson expected a rough passage, but he had not anticipated such scathing attacks on his competence and credibility, and the personal nature of the assault, with the jokes about his accent, left him reeling. He managed to hold his own on technical matters pertaining to locomotives, but he had no defence when the surveying errors were discovered. It was little wonder there were measurement errors when the survey had to be made under the threat of gunfire, but mitigating circumstances held no sway in that legislative arena. His attackers, scenting blood, closed in for the kill, and Stephenson had to endure the humiliation of defeat, followed by the insult of the Rennies being hired above his head.

The only patch of blue in George Stephenson's clouded sky that year was the opening of the Stockton & Darlington Railway. Writing of the event to Robert in Colombia, Joseph Locke reported that 'The opening of the Darlington Railway has made an important impression on the Public which has gained your father much popularity . . .'.

The Rennies carefully prepared plans, sections and estimates of the proposed railway, and the new Bill passed through Parliament the following year with little opposition. Having received parliamentary approval to proceed with the railway, the directors wrote to the Rennies enquiring what terms they would accept to superintend its construction. John Rennie replied that when his brother returned to London within the next few days they would communicate their terms 'explicitly'. At the next meeting of the directors, on 9 June 1826, a letter was read from the Rennies that led to a discussion of 'considerable length'. The celebrated engineers had apparently failed to 'state specifically the terms on which they would undertake the superintendence of the Rail Way'. The directors accordingly agreed 'to avail themselves of the professional assistance either of Mr Rastrick or Mr Stephenson', writing back to the Rennies to enquire whether they were agreeable to working with either party.

George Rennie attended the next directors' meeting in person. After laying down the conditions under which he and his brother were prepared to act as principal engineers, he informed them he had no objection to working with 'Mr Jessop, Mr Telford or any Member of the Society of Engineers . . . but he would not consent to be associated in any way whatever with Mr Rastrick or Mr Stephenson'. Rennie's blunt dismissal of their own engineer seems to have set the directors back on their heels because the meeting was adjourned with no further discussion. Reconvening two days later, they rejected Rennie's terms. A second meeting followed in another two days at which the directors decided to engage a consulting and an operative engineer. The consulting position would be offered to Josiah Jessop, but the decision on the operative engineer was postponed until testimonials had been received on behalf of Messrs Rastrick and Stephenson. Edward Pease wrote a letter supporting Stephenson's candidature, and so did Michael Longridge, another of his business partners. Joseph Sandars made the point that although Rastrick's testimonials were very creditable, it was doubtful whether the terms he was likely to ask would be acceptable. Given the favourable reports supporting Stephenson, Sandars withdrew his nomination of Rastrick, and the directors unanimously resolved to appoint Stephenson as the principal engineer, at an annual salary of £800 (£53,000).

Stephenson was no doubt pleased at his reinstatement, but irked at having his work monitored by Jessop, and from the outset he disagreed with him on almost every issue. He was also obliged to continue working with Vignoles, whose engineering instincts usually coincided with Jessop's. Whenever there was a deadlock, though, the directors always backed Stephenson, undermining Jessop's consultative function. The poor man did not have to endure his insufferable situation for very long because he died, quite suddenly, that October.

Vignoles's position became increasingly difficult, and the final straw came towards the end of that year, when he made some measurement errors in the levels for a tunnel. Recounting his difficulties with Stephenson to a confidant, Vignoles wrote:

> I do also acknowledge having on many occasions differed with him . . . I also plead guilty to having neglected to court Mr Stephenson's favour by sycophantic expressions of praise . . . or by crying down all other engineers, particularly those in London . . .

Vignoles resigned early the following year and was replaced by Joseph Locke, the son of one of Stephenson's old friends, whom he had trained himself.

Although Stephenson had moved back to Newcastle after completing his survey of the line, he still spent precious little time at the Forth Street Works. With Robert still in Colombia, responsibility for the plant had fallen largely to Longridge, but he was busy trying to run his own business, and things inevitably got out of hand. Delivery of locomotives to the Stockton & Darlington Railway had fallen so far behind schedule that *Locomotion* was the only engine on the line for the entire inaugural year. And when engines were eventually delivered they were frequently found to be faulty. *Hope*, the railway's second locomotive was delivered towards the end of 1825, and could not be made to work at all. The beleaguered Longridge wrote to Richardson, hoping 'that Robert's early return to England will soon relieve me . . .'. He concluded with the plea that he would be obliged 'if you or Mr Pease can appoint a more suitable person . . .'.

Robert Stephenson was frequently reminded of the plight of the locomotive factory. 'I know not how the Manufacturing goes on at Newcastle,' Locke wrote to him, early in 1827, '[but] I fear, not so briskly as it has done – I believe Mr Longridge wishes to declare the Engine Works closed until your return . . .'. Robert also received a gloomy letter from Edward Pease, telling him how much the locomotive business was suffering through his absence. The Stockton & Darlington Railway was desperately short of locomotives, but the Forth Street Works was unable to satisfy their needs. No other suppliers could be found, and the whole future of the locomotive was looking rather bleak. Indeed, the Hetton Colliery had already began replacing some of its Stephenson-built locomotives with stationary engines, and the newly formed Newcastle & Carlisle Railway rejected steam power in favour of horses.

The locomotives being built at the Forth Street Works at this time – albeit slowly – were little advanced over the ones George Stephenson had built during the early days at

Killingworth. Improvements had been made in the way the wheels were cast, and they were now fitted with wrought-iron tyres which could be replaced as they wore down. The front and back wheels were also linked together with side coupling rods instead of chains, an improvement apparently suggested by Hackworth, though there have been other claimants taking credit for the device. Experiments had also been conducted in pre-heating the feed water to the boiler with exhaust steam, and in using water-filled tubes at the bottom of the fire grate to increase the total heating area, but with little success. Regardless of these minor improvements, the engines were severely handicapped by the inadequate steaming capacity of their boilers, which all had straight-through flues. This was not such a disadvantage on the short-haul colliery lines because locomotives could build up a head of steam during their long stops at either end. But it was a serious deficiency on long runs, like the Stockton & Darlington Railway, where locomotives sometimes ran out of steam. Hackworth's *Royal George*, with its return-flue boiler, was a significant improvement, having about a threefold increase in heating area, but this was not a radical advance. What was needed was a fundamental change in locomotive design, but there was not an engineer in the land capable of effecting such a revolution. The sorry state of the locomotive at that time may explain why Nicholas Wood, of all people, should have predicted:

> It is far from my wish to promulgate to the world that the *ridiculous* expectations . . . of the enthusiastic specialist will be realised, and that we shall see engines travelling at the rate of *twelve*, *sixteen*, *eighteen*, or *twenty* miles an hour. Nothing could do more harm towards their adoption or general improvement than the promulgation of such *nonsense*.[20]

The downturn in the fortunes of the Stockton & Darlington Railway during this period is reflected in the reminiscences of Edward Pease, the railway's first director.

> My favoured position in life [as director] did not render any remuneration for service needful, nor did I ever receive a shilling . . . for my exertions. When towards the close of our work, money falling short, our banker refused to grant us more money, I paid all the workmen . . . employed in this way out of my own resources . . . I remained in the direction . . . until 1827 . . . I retired with a resolution never to enter a railway meeting again!

Robert Stephenson's last South American letter to Longridge, written 16 July 1827, began jubilantly: 'The period of my departure from this place has at last really and truly arrived . . .'. He went on to say he had recently received a letter from Richardson stating 'that the factory was far from being in good condition and that unless I returned promptly to England it would not improbably be abandoned . . .'. He had hoped to visit the Isthmus of Panama on his way home, 'so I may know something about the possibility, or impossibility, of forming a communication between the two seas . . .'. However, his concern over the delay such a trip would cost caused him to change his plans, and he proceeded straight away to Cartagena, where he hoped to find a direct passage to England. Stephenson was joined there by one of the employees of the Mining Association, a man named Gerard who was bound for Scotland. And it was there that a most remarkable coincidence took place.

Two men, both tall, converse together in the public lounge of an inn. They speak in English. One wears a broad-brimmed hat, his weather-beaten face contrasting starkly with the pale straw. He probably looks older than his fifty-six years. He has clear blue eyes. But the most striking feature of the man is his incessant pacing – from one end of the room to the other, like a caged panther. What makes him so restless?

Two other men, Stephenson and Gerard, share the same room. Their conversation together is less animated. They too are speaking in English, and when the pacer hears the language of his birth he strides over to join the conversation. Suddenly Stephenson and the pacer, to each one's profound amazement, learn the identity of the other.

This is one version of the historic meeting between Stephenson and Trevithick, the version that Stephenson remembered. The other, given by one James Fairbairn, was corroborated by Trevithick's companion in the Colombian inn. That companion, an Englishman named Bruce Napier Hall, was then a serving officer in the Venezuelan army. And it was in a Venezuelan jungle, just a short time before, that he had saved Trevithick's life. According to Fairbairn:

> Mr Stephenson . . . like most Englishmen, was reserved, and took no notice of Mr Trevithick, until the officer said to him, meeting Mr Stephenson at the door, 'I suppose the old proverb of "two of a trade cannot agree" is true, by the way you keep aloof from your brother chip. It is not thus your father would have treated that worthy man, and it is not creditable to your father's son that you should be here day after day like two strange cats in a garret; it would not sound well at home.' 'Who is it?' said Mr Stephenson. 'The inventor of the locomotive, your father's friend and fellow worker; his name is Trevithick, you may have heard it,' the officer said; and then Mr Stephenson went up to Trevithick. That Mr Trevithick

> felt the previous neglect was clear. He had sat with Robert on his knee many a night while talking to his father, and it was through him Robert was made engineer. My informant states that there was not that cordiality between them he would have wished to see at Cartagena.

The officer confirmed the lack of cordiality, adding that:

> it was quite possible Mr R Stephenson had forgotten Mr Trevithick, but they must have seen each other many times. This was shown by Mr Trevithick's exclamation, 'Is that Bobby?' and after a pause he added, 'I've nursed him many a time.'

It is difficult to rationalise Stephenson's apparent coolness towards Trevithick, especially given his compassionate nature. Was he merely acting out of deference to his father's disdain for perceived rivals? This seems unlikely because Robert was always showing kindnesses to men, like James Walker, whom his father had cast aside. It is equally implausible that Robert's unfortunate experiences with the unruly miners in Colombia had sensitised him against all Cornishmen. The reason for his attitude towards Trevithick remains a mystery.

Trevithick's rescue by Bruce Napier Hall, and the events leading up to it, might have been taken from the pages of a boys' adventure story of yesteryear. Following his departure from Peru, Trevithick spent some time mining in Costa Rica. Presumably the winnings were not rich enough to induce him to stay, and he left in search of a new route from the Pacific to the Caribbean coast, across the mountains. The excursion, which lasted about three weeks, took his small party through woods and swamps and across rapids. They lived on a diet of wild fruit, and the occasional monkey they chanced to shoot with the single fowling piece they carried. While making a crossing by raft, Trevithick and two of his companions got stranded on one side of a river while the rest of the party reached the other side safely. Without another alternative, the three men had to swim across the river. One of them drowned and Trevithick only just escaped with his life. They finally reached the Caribbean coast, arriving in a state of great exhaustion.

On his last adventure, in Venezuela, Trevithick offended one of the locals, who retaliated by capsizing his boat in the River Magdalena. Fortunately for the Cornishman, Bruce Napier Hall was on the bank, shooting wild pigs.

> He heard Trevithick's cries for help, and seeing a large alligator approaching him, shot him in the eye, and then, as he had no boat, lassoed Mr Trevithick and . . . drew him ashore much exhausted and all but dead. After doing all he could to restore him, he took him on to Cartagena . . .

Trevithick was destitute, and if it had not been for Hall he would not have reached Cartagena. And, regardless of Stephenson's seeming inhospitality towards him, he did give him the money for his fare home.

Trevithick 'had a very good passage home, six days from Carthagena [*sic*] to Jamaica, and thirty-four days from thence for England . . .'. Stephenson and Gerard were less fortunate. Unable to obtain a direct passage to England, they boarded a ship bound for New York. It was to be an eventful voyage, as Stephenson recounted to a friend the following year:

> We had very little foul weather, and were several days becalmed amongst the islands; which so far was extremely fortunate, for a few degrees farther north the most tremendous gales were blowing; and they appear . . . to have wrecked every vessel exposed to their violence, of which we had two appalling examples as we sailed north. We took on board the wrecks of two crews who were floating about in dismantled hulls. The one had been nine days without food of any kind, except the carcasses of two of their companions who had died a day or two previous . . . To attempt any description of my feelings on witnessing such a scene would be useless. You will not be surprised to know that I felt somewhat uneasy when I recollected that I was so far from England, and that we might be wrecked.

Their ship made a safe passage to North America and the voyage was almost at an end when they were hit by a hurricane. With the wind screaming in the rigging the ship struck rocks close to shore and quickly began taking on water. The vessel began breaking up, but the sea was running too high to launch the lifeboats and the desperate passengers that crowded the tilting deck were convinced they were all going to perish. It was a long and terrifying night, but dawn broke to calm seas. Miraculously everyone was rescued, without the loss of a single life. Most of their belongings were lost, including Stephenson's trunk containing his money, but, fortunately for him and his travelling companion, he was able to obtain funds in New York. Stephenson decided to see a little of North America before venturing to sea again.

What passed between Stephenson and Trevithick during the time between their meeting in Cartagena and their departure for home is a matter of speculation. In preparing his (1866) biography of Robert Stephenson, J. Cordy Jeaffreson had access to 'a mass of evidence' that was apparently unavailable to other biographers. Without citing any references, he claimed that 'There is no doubt that the original and daring views of Trevithick with respect to the capabilities of the locomotive made a deep impression on Robert Stephenson.' Regardless of the veracity of Jeaffreson's claim, when Stephenson returned to England he set about the task of effecting major improvements in the locomotive. Some idea of his thoughts on the matter is contained in a letter he wrote to Longridge on New Year's Day 1828, barely two months after his return:

> Since I came down from London, I have been talking a great deal to my father about endeavouring to reduce the size and ugliness of our travelling engines, by applying the engine [cylinder] either on the side of the boiler or beneath it entirely, somewhat similar to Gurney's steam coach. He has agreed to an alteration which I think will considerably reduce the quantity of machinery as well as the liability of mismanagement. Mr Jos. Pease writes my father that in their present complicated state they cannot be managed by 'fools', therefore they must undergo some alteration or amendment. It is very true that the locomotive engine, or any other kind of engine, may be shaken to pieces; but such accidents are in great measure under the control of enginemen, which are, by the by, not the most manageable class of beings. They perhaps want improvement as much as the engines.

Robert Stephenson, who had left England before celebrating his twenty-first birthday, returned as a twenty-four-year-old man of the world, eager to take the steam locomotive to its next level of development.

CHAPTER 7

Rocket on trial

Thursday 8 October 1829, the third day of the trials, was another cool autumnal day, following a night when temperatures dipped to within a few degrees of freezing. The mercury would not climb very much higher during the day, but at least yesterday's storm clouds had cleared away and it promised to be fair. Some of the early spectators may have been tempted to partake of a hot toddy, but most gentlemen would have been content to sip their coffee, and perhaps peruse the morning papers.

The lead article in *The Times* confirmed the signing of a peace treaty between Turkey and Russia, following a year of war. Readers were warned that Russia would exact 'rigid terms' from the vanquished Turks. This article was followed by a smuggling story about contraband despatched to Lord Stuart, the British ambassador in Paris. Allegations in the French press claimed his lordship had used his influence to prevent the confiscation of the illicit goods. *The Times* stopped short of pre-judging the case, but it gave no quarter to another peer, the Duke of Newcastle, who had evicted tenants from his land for not voting for his chosen candidate. His argument, 'Am I not to do what I will with my own?' so incensed that bastion of morality that it wrote, 'The Duke of Newcastle himself may not have brains enough to see the full bearing of his own manoeuvres, but there are other "borough-mongers" who can.' Under the headline 'MELANCHOLY OCCURRENCE' was the sad tale of a respected young lady, rejected by her feckless fiancé. His excuse that his aunt had forbidden the marriage was patently untrue, and when she discovered he had married another she took her life with 'a pennyworth of arsenic'.

Theatregoers would have learned that Miss Fanny Kemble was starring in *Romeo and Juliet*, at the Theatre Royal, in Covent Garden. Just two evenings before, the vivacious young nineteen-year-old had made her stage debut, alongside her father, the producer, who played Mercutio, and her mother, in the role of Lady Capulet. Fanny Kemble had been a sensation, not only for her superb performance but also for debuting to help save her father from financial ruin. Charles Kemble, like his wife, Marie, was a celebrated actor, but his theatre had fallen on hard times and was £13,000 (£760,000) in debt. His daughter was an instant success, and the fortunes of the theatre dramatically recovered as audiences flocked to Covent Garden to see the new sensation. Her beauty and intelligence were matched by a sparkling personality, making her the object of attention of many admirers. George Stephenson himself would fall under her spell, and she, in her turn, would be enchanted by him.

Newspapers were not the only items of print circulating the course that morning. Before the competition got under way cards were distributed setting out the rules and conditions of the competition. These were the same rules the judges had drawn up at the end of the first day, and were therefore dated 6 October. It was most unfortunate the cards had not been available earlier because it gave some people the false impression that the judges had somehow changed the rules midway through the competition. According to *The Liverpool Chronicle*:

> The object of giving it the date of the 6th is to make it appear as if it had been concocted on the day of the competition, and before it commenced; whereas the fact is, it was never seen or heard of . . . till the morning of the 8th, two days later.

The editorial pointed to discrepancies between the new rules and the original stipulations, including the omission of the requirement that locomotives must consume their own smoke, and asked which locomotive benefited most. The reader was led to the inevitable conclusion that *Rocket* was the beneficiary, because it had made smoke. If the newspaper had been fair it would have reported that this occurred only on the first day, when some coal had been inadvertently mixed with the coke. But *The Liverpool Chronicle* was clearly not interested in giving an honest report of the trials – its object was to insinuate that the judges were biased in favour of the Stephensons.

In accordance with the rules, *Rocket* was weighed, 'with its full complement of water in the boiler' and with 'no fuel in the fireplace'. *Rocket* was too long to fit on the scales, so they had to weigh the back end first, then the front, adding the two readings. She weighed in at four tons and five hundredweight (cwt), which was 560 lb under the stipulated maximum of four and a half tons. The assigned load of three times the locomotive's weight was accordingly set at 12 tons

15 cwt. This comprised two carriages loaded with stones, together with the tender and its complement of water and fuel, which counted as part of the load. Once the tender and the two carriages had been coupled to the engine, the entire train was moved up to the starting line, at the extreme west end of the course. A good deal of muscle and sinew was needed to set eighteen tons of iron and stone into motion, but the company had many workers on hand for such tasks.

The time had come to lay the fire in the firebox and see how long it took, and how much coke was used, to raise steam. Access was by a small hinged door at the back of the engine, held closed by a latch. Once the fireman had laid the grate with kindling and built up a small pile of coke, he lit it and closed the door.

Like many other locomotives of her time, *Rocket* did not have a pan beneath the grate to catch the ash, consequently hot cinders fell on to the track when she was in operation, posing a fire hazard. Later locomotives were fitted with ash-pans, and also with dampers to control the flow of air through the bottom of the fireplace. The fire in the grate was in continuity with the chimney, and once it got started and hot gases began passing up the chimney, forming a draught, the fire began to draw. This pulled fresh air through the bottom of the grate, making the coke glow red hot. Each time the fireman opened the door to add more coke the incoming air took the path of least resistance, entering the firebox through the door rather than through the bottom of the grate. This stopped the fire from drawing, and if he had tarried the fire would have died down. But once he closed the door the fire began drawing again. When a locomotive was kept stationary for any length of time it was recommended that the firebox door should be left open to stop the fire from drawing. Keeping the fire hot, but not roaring, saved fuel and reduced steam pressure, so preventing the wasteful venting of excess steam through the safety valve.

According to the company's original stipulations, all locomotives had to be fitted with two safety valves, one of which must be inaccessible to the crew. There also had to be a mercury pressure gauge. *Rocket*'s accessible safety valve was of the simple weighted lever type, widely used on stationary engines since before the time of locomotives. A lever passed over the top of a valve and the force pressing down upon it could be increased by adding more weights to the lever's free end. By adjusting the weight the valve could be set to vent at any particular steam pressure. This was set at 50 psi for the trials. While simple and effective, the valve had the disadvantage that the lever tended to bounce when the locomotive was in motion, causing intermittent venting. It was also easily tampered with by crews, who could increase steam pressure simply by tying the lever down. An effective alternative was the spring-loaded valve, which eliminated intermittent venting. This had been in use by Blenkinsop since as early as 1812, but it was not very reliable, which prompted Hackworth to invent an improved version.

Rocket's inaccessible safety valve was housed inside a padlocked casing, located towards the front of the boiler, behind the chimney. There is little information about it, but it was almost certainly spring-loaded. Its housing was a simple cylinder, and this has often been misidentified as a dome. But a dome is an entirely different structure with a different function, and *Rocket* was not so equipped at the time of Rainhill.

Domes, not introduced until 1830 and still used in modern locomotives, were devised to minimise 'priming', where water droplets get picked up by the steam on its way to the cylinders. This is undesirable because it introduces unwanted water into the cylinders, hampering their function and potentially causing damage because water is incompressible. Priming could be reduced by placing the supply pipe for the regulator valve as high as possible inside the boiler, but some droplets inevitably got picked up from the boiling water below. Adding a dome to the top of the cylinder and bending the supply pipe to reach high inside it essentially eliminated the problem.

The mercury gauge was a simple U-tube filled with mercury, one of its limbs being attached to the boiler and the other open to the atmosphere. This type of pressure-measuring device is called a manometer, and is typically made of glass. As the steam pressure increases the mercury is forced higher up the open-ended limb while being depressed in the other one. The difference in levels between the two, viewed through their glass walls, is read off as pressure. *Rocket*'s gauge was made of copper pipe, so it could not be read like a glass manometer. Instead, the open-ended limb was fitted with a wooden float – presumably a calibrated rod that projected above the copper pipe, enabling the driver to read the pressure.

Pressure is always measured *relative* to atmospheric pressure, which is about 15 psi. A boiler pressure of, say, 50 psi means the pressure inside the boiler is 50 psi *above* atmospheric pressure. A boiler pressure of 50 psi would cause a difference in mercury levels of about 100 in. The pressure could therefore be expressed as '100 inches of mercury.' To accommodate peak pressure above 50 psi *Rocket*'s mercury gauge had to be about 9 ft tall, and was braced against the side of the chimney. Its primary role was to give accurate pressure readings, but it also functioned as an additional safety valve, because if the two valves failed and the pressure rose dangerously high, the mercury would be ejected from its open limb, releasing the steam.

As the judges recorded in their notebooks, *Rocket* took fifty-seven minutes to reach a boiler pressure of 50 psi, consuming 142 lb of coke in the process. But her crew did not sit idly by while they waited for the pressure to rise: steam locomotives are demanding creatures, requiring constant care and attention. The crew had probably been hard at work long before the judges arrived. The locomotive had been in steam the previous day and they had probably already raked out any residual ash from the firebox, and clinker from the grate, the night before. Clinker is the intractable mass that solidifies from molten slag when fuel is burned at high temperatures. If it is not broken up and removed it soon clogs the bars, stifling the draught and hampering steam production. Clinker must have been a serious problem for the Rainhill competitors, just as it was for the crews in the re-enactment of the trials at Llangollen, North Wales, in 2002. *Rocket*'s crew may also have checked the base of the chimney for accumulated soot, though it is unclear whether their engine had an inspection panel for that purpose at the time of the trials.

In between tending his furnace, the fireman, like the driver, continued his rounds with the oilcan. Aside from applying a squirt of animal or vegetable oil to each of the oiling points, oil was applied liberally to every moving surface, including the piston rod with its attached cross-head, the slide-bars upon which it slid, and the bearings at either end of the connecting rod. Every one of the moving joints in the valve gear linkage had to be lubricated, together with the valves themselves. Then there was the feed pump, the regulator valve, the taps on the try cocks for checking water levels, the leaf springs of the wheels, and every other place where metal moved against metal, including the hinges and latch on the firebox door. The upper end of each cylinder had an oil reservoir which had to be checked and topped up, and this required regular attention when the engine was running. While the driver did his final checks with a spanner, making sure all the vital nuts were good and tight, his fireman probably busied himself with cleaning. All bare metal surfaces were wiped down with an oily rag, while the paintwork and copper pipes were made to gleam with clean cloths and elbow grease.

Having set their watches to precisely the same time, Messrs Rastrick and Wood took their places along the track. John Rastrick positioned himself at post No. 1, 220 yards downline from the start and still visible, if not recognisable, to *Rocket*'s crew. Nicholas Wood, a mile and a half further on at post No. 2, was completely out of their sight. The judges were ready. *Rocket* was ready. The signal was given and the trial began.

Rocket was an innovative locomotive but she did have antecedents. Her ancestor was *Lancashire Witch*, initially called the 'Liverpool Travelling Engine', built at the start of 1828, less than two months after Robert's return from Colombia. In the original design the boiler was to have three parallel flue tubes running the entire length of the boiler barrel, thereby increasing the total heating area. The central tube was to be the broadest, flanked by narrower ones. The idea for this new boiler came from Henry Booth. That the company treasurer should have been contemplating boiler designs is quite remarkable, but Booth had always been interested in mechanical engineering, and if he had not entered the world of commerce he would doubtless have become an engineer. Booth's principal objective was to devise a boiler that would burn coke, thereby complying with the government's stipulation that locomotives must be smokeless. To this end the primary draught for the furnace was to be provided by a pair of bellows, and the blast from the exhaust steam was to be ancillary: a strong air flow was necessary because coke burns far less readily than coal.

Although George Stephenson carried out some initial experiments to test the feasibility of the new boiler design, it was left to his son, and the skilled men at the Forth Street Works, to build it. The design was subsequently changed to two flue tubes of equal width, lying side by side and joined to a single chimney. This was essentially a return-flue boiler, but the chimney was at the opposite end to the fire, and each flue tube had its own fire grate. Joining the two flue tubes to the chimney was not an easy task, as George conceded in a letter to Robert:

> I am quite aware that the bent tubes are a complicated job to make, but after once in and well done it cannot be any complication in the working of the engine. This bent tube is a child of your own, which you stated to me in a former letter . . .

Apart from showing George's lack of sympathy for his son's manufacturing problems, the letter reveals that Robert was just as involved in the design of *Lancashire Witch* as he was in its construction. Indeed, Robert seems to have taken a leading role, and it was on his own initiative that he placed the cylinders on the outside of the boiler. He inclined them at an oblique angle, to reduce the rolling effect of vertically aligned pistons. No other locomotive had inclined cylinders, and these were mounted at the back of the engine, with their pistons connected to the front wheels, by

Lancashire Witch.

crankpins. For lightness, all four wheels were made of wood, fitted with wrought-iron tyres. All these features were adopted in *Rocket*, giving the two locomotives a striking superficial resemblance. Like Hackworth's *Royal George*, the wheels were fitted with leaf springs, reducing the stresses on rails and wheels, and giving a better ride. Steel plate suitable for making sufficiently strong springs had been available only since about 1825, apparently pioneered by Nicholas Wood, who was probably also the first to adopt wrought-iron tyres.

Lancashire Witch's most advanced feature was that the pistons worked expansively, meaning that the steam supply to the cylinder was cut off before the piston had been pushed all the way to the end of the cylinder. Modern steam locomotives operate expansively, and the cut-off point for the injection of steam can be varied according to operational circumstances. When the engine is starting off, maximum force is required to overcome the inertia of the train. The inlet port for the steam is therefore kept open for the entire piston stroke, so the full force of the steam pushes against the piston during its complete traverse of the cylinder. During this time the steam pressure inside the cylinder is maximal, being almost the same as that inside the boiler. Once the locomotive is moving it no longer requires the maximum force to act upon the piston for the full stroke, so the steam is cut off well before the piston completely traverses the bore.

At the precise moment of cut-off the steam pressure inside the cylinder is maximal, and therefore exerts the greatest force on the piston. But as the piston continues moving down the cylinder the volume available to the trapped steam continually increases, and the steam expands to fill all the available space. As the steam expands it continues exerting a force on the piston, and this has been described as the expansive property of steam. But as the steam expands its pressure progressively falls, so the force it exerts on the piston continually decreases.

The rationale for running pistons expansively is to save steam, and therefore fuel, and a skilled driver of a modern locomotive can operate his engine with great economy. At the start of the journey steam is admitted into the cylinder for maybe 70 per cent of the piston stroke. As speed is gathered the percentage is reduced, and, once the train has reached its cruising speed, the steam may be admitted for only about 5 per cent of the stroke. Modern locomotives have a variable cut-off by virtue of the sophisticated gearing of the valve mechanism, but the driver of *Lancashire Witch* had only two choices: cutting off the steam at 50 per cent of the stroke, or not cutting it off at all. *Rocket* was not fitted with expansive valve gear. The original valves have not been preserved, so we cannot be sure, but it seems likely they would have been set to deliver steam for most of the stroke, to exert maximum force. This would have been ideal for the frequent stops and starts of the Rainhill trials. *Rocket* was not designed for efficiency: she was designed for winning the contest.

Mr Rastrick stares intently at his timepiece as the locomotive clatters and chuffs past his post for the very first time, catching the instant of passing in his peripheral vision. It is ten-thirty-eight and fifteen seconds precisely. The time is dutifully recorded in his notebook, alongside the starting time of ten-thirty-six and fifty seconds. From a standing start it has taken *Rocket* one minute and twenty-five seconds to cover the 220 yard distance to the first post.

From Mr Wood's perspective the engine is just a dot in a haze of steam. But in seven minutes and forty-three seconds precisely this has transformed itself into a passing train. Two minutes and fourteen seconds later *Rocket*, running backwards, steams past Mr Wood for the second time, having coasted to a halt at the finish line 220 yards away, changed directions and charged back again in reverse. The arithmetic could be done later – all he has to do for now is record the times. But if he had done the calculations he would have found that *Rocket* completed her first one-and-a-half-mile leg at 'full speed' at 11.67 mph. This was no great speed, but it was above the stipulated minimum of 10 mph. The return leg is a little faster: 13.43 mph.

Operating a locomotive like *Rocket* required intuition as well as skill. The mercury pressure gauge was primarily to satisfy the stipulations of the company, and the crew were not used to such refinements. But they might have found it very useful under racing conditions to help them keep the boiler pressure up to maximum, which was the fireman's responsibility. Without a gauge the fireman had to use his sixth sense to know when the steam pressure was falling, and how much coke to shovel to maintain the pressure. He also had to monitor the water level in the boiler, and open the valve of the feed pump when more was needed. The driver would have kept a watchful eye on the water level too, because of the dire consequences of its falling too low.

One of the driver's most important skills was knowing when to reverse the valve sequence in readiness for the return leg of the lap as precious moments could be lost during each turnround. The frequent changes in direction put a great deal of reliance on the valve-changing mechanism. Anticipating this, Robert Stephenson had paid a great deal of attention to the reversing mechanism. 'I expect the mode for changing the gear will please you,' he wrote to Booth six weeks before the trials. 'It is now so simple as I can make it and I believe effectual—'

Rocket, running forward now, passes Mr Rastrick again and approaches Mr Wood for the third time. And as Mr Wood stares intently at his watch, *Rocket*'s driver holds the handle of the regulator valve, biding his time. As judge and locomotive cross and part, the driver looks ahead towards the finish line, and when the time is right he swings the regulator handle fully to one side, cutting off steam to the cylinders. The locomotive immediately begins to slow. He could slow down faster if he reversed the valve sequence, effectively changing into reverse gear, but he cannot attempt this manoeuvre until the locomotive is going much slower – less than about 8 mph – otherwise he could damage the equipment.

Judging his moment carefully, the driver stamps down firmly on a small round foot pedal that protrudes a few inches above the level of the footplate. He has to use most of his weight because the pedal is spring-loaded and has considerable resistance. The shaft is notched, a little way below the pedal, and as it is forced down through its hole the notch comes up hard against the underside of the footplate, locking the pedal in the down position. His action has no apparent effect, and *Rocket* continues her majestic loss of momentum. But the act of depressing the pedal has reversed the action of the valves supplying steam to the two cylinders. Details of how this mechanism works are given in the next four paragraphs, which can be skipped by those not interested in such technicalities.

The forward-and-reverse mechanism is mounted on the front axle, just to the left of centre. The mechanism, which is attached to a short sleeve through which the axle rotates, includes a pair of eccentrics. The two eccentrics are sandwiched between a pair of discs, called cheek plates.

As the eccentrics rotate, their embracing straps oscillate,

and these movements are transmitted to a pair of rods, called eccentric rods, to which they are attached. The eccentric rods are linked by a series of rods and shafts to the cylinder valves. One eccentric operates the left cylinder valve, the other the right one. On either side of the eccentric assembly is a collar, which is clamped to the axle. The two collars, called drivers, are like a pair of bookends for the eccentric assembly. But they are not a tight fit because the entire eccentric assembly can be moved about two inches from side to side, by the slippage of the short sleeve upon the axle. When the assembly is moved to the left the left-hand cheek plate presses against the left-hand driver while the right-hand cheek plate loses contact with the right-hand driver. The opposite happens when the eccentric assembly is moved to the right. But there is more than mere touch contact between cheek plate and driver because the driver has a rectangular projection, called a dog, which fits into an elongated slot in the cheek plate.

If the eccentric assembly is disengaged from the right-hand driver and pushed up against the left-hand one, the dog runs around the cheek plate until it locates the slot. Provided the locomotive is moving slowly enough the dog will then slip into the slot, joining the eccentric assembly to the driver and locking the locomotive in forward mode. When locked in this position the reciprocating movements of the eccentrics operate the cylinder valves so that the pistons make the wheels rotate forward. The left-hand driver, with its dog, is therefore for moving forward. To make the pistons work in the opposite direction and drive the wheels backward requires the valves to be in different phases. This is achieved by pushing the eccentric assembly into contact with the right-hand driver because its dog is out of phase with that of the left-hand driver. The sideways movements

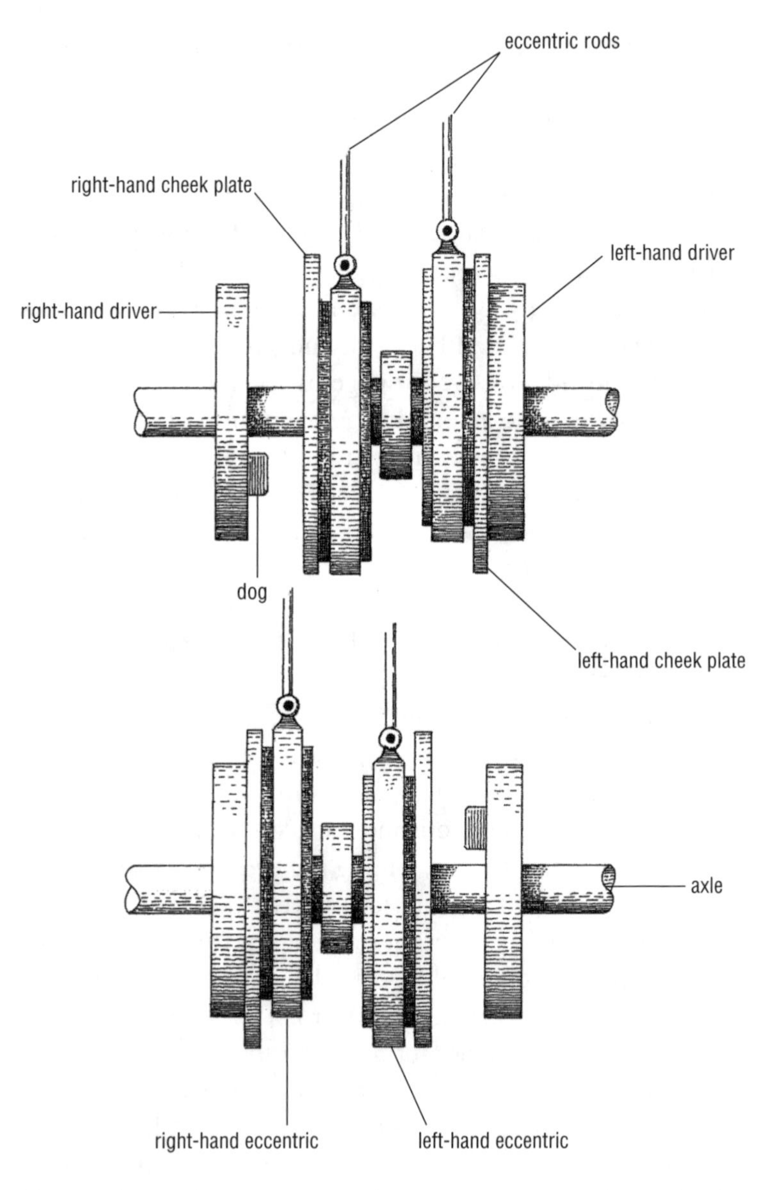

Rocket's forward-and-reverse mechanism, seen from the front, showing the eccentric assembly moved to the left (top), and to the right.

of the eccentric assembly are controlled by a Y-shaped bracket or yoke. The yoke is attached to a shaft which lies parallel with, and just below, the front axle.

The reversing pedal is attached to a side arm beneath the footplate which connects to a long rod, the reversing rod, running beneath the locomotive, to the yoke shaft to which it is connected. When the reversing pedal is depressed the reversing rod rotates clockwise. This pushes the yoke towards the right, shifting the eccentric assembly in the same direction to lock against the right-hand driver, so reversing the action of the valves to turn the wheels backward. Forward motion is effected by delivering a sideways kick to the reversing pedal. This jogs the notched shaft free from its locked position, causing the pedal to spring up. As it does so, the reversing rod rotates anticlockwise, causing the yoke to push the eccentric assembly to the left, locking it against the left-hand driver to change the valve action to forward motion. The mechanism is unsophisticated, almost crude, but it worked, and worked well.

If *Rocket*'s driver is not careful he will lose valuable time during the turnarounds. He needs to reach the finishing point as quickly as possible, bring the engine to a halt, and then head off again in the opposite direction. Just seconds earlier he stomped down on the reversing pedal, reversing the valve sequence to make the engine run backward. If he keeps the steam cut off he will continue slowing down gently as the locomotive coasts forward. But he runs the risk of overshooting the finish line, and he knows he has the additional momentum of the twelve-ton load pushing from behind. If he applies steam he will slow down more quickly because the pistons will be opposing forward motion in trying to drive the engine backward. But if he applies steam too early, or with too much force, he will slow down prematurely and undershoot the finish line. In that event he will have to uncouple the valves from the eccentrics and operate them by hand, costing valuable time. It is all a matter of judgement and timing.

The skills of the driver and fireman were only part of the equation. The vital other part was the locomotive itself. *Rocket* was fundamentally different from anything that had ever rolled out of the Forth Street Works. But it had to be a fundamentally new engine design if it were to win the competition.

When Messrs Walker and Rastrick were on their fact-finding mission the previous year to help the Liverpool & Manchester Railway Company decide between fixed and locomotive engines, they saw an experiment with *Lancashire Witch*. Hauling a load of forty tons she achieved a speed of almost 9 mph, while ascending a slight incline. *Royal George* did somewhat better, but on a slightly lesser gradient. *Lancashire Witch* was undoubtedly an excellent engine, the best the Stephensons had ever built, but, like *Royal George*, her steaming capacity was only adequate. That is why *Lancashire Witch*, like her contemporaries, attained such modest speeds. Without some revolutionary new development in boiler design the steam locomotive would never reach its full potential.

Soon after the announcement of the Rainhill trials Henry Booth approached George Stephenson with a proposal. He had thought of an entirely new way of increasing steam production. Recalling the event some years later, Booth wrote:

> I mentioned my scheme to Mr [George] Stephenson, and asked him if he would join me in building a locomotive to compete for the prize . . . Mr Stephenson took a day or

> two to look into the merits of the plan I proposed, and then told me he thought it would do, and would join me in the venture.

The key to overcoming the limited steaming capacity of the boiler was to devise a way of radically increasing the surface area available for heat exchange between the furnace and the water. All previous attempts to do this had been woefully inadequate. Return-flue boilers had increased surface areas beyond those of straight-through flues, but not significantly. Stephenson's original plan to fit *Lancashire Witch* with three flues would have taken this a step further, and some boiler designers thought it was the logical direction in which to go. There had been several proposals to build boilers with multiple flues, and many years earlier Trevithick had received a letter from a man proposing a boiler with three banks of triple flues. Even if the boiler had been built, it is doubtful it would have succeeded, partly because of the poor circulation of heat between the hot flues and cold water.

An entirely different approach was to pass water through a number of small tubes inside the furnace. George Stephenson had built two such locomotives for the St Etienne Railway in France, but as Robert Stephenson wrote: 'the expedient was not successful; the tubes became furred with deposit, and burned out.'

Booth's idea was for a multi-tubular boiler where a large number of 3 in.-diameter copper tubes would carry hot combustion gases from furnace to chimney, passing through the water in the boiler barrel. The advantage of using a large number of small tubes, rather than a few large ones, is that the surface area for heat exchange is greatly increased. This is because of the relationship between size and surface area:

as things get smaller their surface area becomes relatively larger. This fundamental relationship between size and surface area, which has far-reaching consequences in both the animate and the inanimate worlds, explains why hot peas on a plate cool down much faster than a hot potato.

As Booth explained, by using multiple tubes:

> we not only obtain a very much larger heating surface, but the heating surface is much more effective, as there intervenes between the fire and the water only a thin sheet of copper or brass, not one-inch thick, instead of a plate of iron of four times the substance, as well as an inferior conductor of heat.

Here was a revolutionary new way of emancipating the locomotive from its evolutionary straitjacket. If Booth's innovative thoughts could be transformed into metallic reality, they would have a locomotive to be reckoned with. But there were formidable technical problems to overcome in building such a sophisticated boiler, and they had less than six months in which to resolve them and construct the entire locomotive.

The concept of a multi-tubular boiler was something entirely new to them, but the very same idea had already occurred to somebody else. Indeed, a patent for such a boiler had been applied for more that two years earlier. Fortunately the inventor, a French engineer named Marc Séguin, was not a contender for the locomotive prize, and his multi-tubular locomotive would not be tested until two months after Rainhill anyway.

Robert Stephenson modestly wrote that 'It was in conjunction with Mr Booth that my father constructed the "Rocket" engine.' The venture certainly began as a joint project between Henry Booth and George Stephenson, and Robert's name was not officially added to the partnership until shortly before the trials began. However, the 'Premium Engine', as *Rocket* was first known, was very much Robert's locomotive, even though the original concept for the boiler was Booth's, and George Stephenson's practical advice was important during construction. He was entirely responsible for its design and manufacture, including such features as the simplified reversing mechanism and the unique firebox, though he drew freely upon the skills of his experienced staff at Forth Street. Foremost among these was George Phipps, his draughtsman, and William Hutchinson, the works manager, both of whom played a major role in the design and construction. 'Mr Stephenson was always ready to avail himself of the abilities of those around him,' wrote Phipps, fifty years after the event, 'his kindly consideration always eliciting the best fruits of their powers.'

Robert Stephenson had drawings made before any construction began, and it was Phipps's job to draft these:

> Having made the original drawings under Mr Robert Stephenson, I can bear witness to the care and judgment bestowed by him upon every detail. The arrangement of the tubes with the method of securing their extremities, the detached fire-box, and many other matters of detail, all requiring much consideration. Mr Stephenson was well aided in all the mechanical details by the late Mr William Hutchinson.

Hutchinson, who eventually became a partner in the business, was referred to as 'the oracle' and when Stephenson encountered a particularly tricky problem he would shake his head and goodnaturedly call out, 'Come, this is a touchy point, let's call in "the oracle".' The biggest problem during *Rocket*'s construction was fitting the copper tubes to the boiler barrel so they would not leak. Twenty-five tubes had to be fitted, and there must have been times when Robert wondered whether he would ever succeed. The supplier had made each tube from a narrow rectangle of copper sheet, 3/32 in. thick, by forming it over a mandrel and brazing together the join.

Great care had been taken in the selection of the wrought-iron plates for the boiler, which was 6 ft long and just over 3 ft in diameter. Boiler plate was now available in large pieces, and only four were needed for the body of the barrel. Each one was approximately 5 ft long and 3 ft wide. The two ends of the boiler were flat, each formed of a single plate.

If the boiler had been built with a straight-through flue the lower half of each end-plate would have been perforated by a large round hole for the attachment of the flue tube. Riveting a flue tube in place so the joints did not leak would have required skill, but was well within the capacity of a good riveter. But *Rocket*'s end-plates were to be perforated by twenty-five holes, and nobody had ever attempted fitting copper tubes into a boiler before. How to do it?

According to Samuel Smiles, Stephenson first tried soldering them to brass screws, which were then screwed into the end-plates – more properly called tube-plates. However, when the boiler was filled with water and pressurised with a pump it leaked so badly that the factory floor was flooded. When Robert reported the failure to his father he wrote back suggesting he soldered the tubes direct to the tube-plates. This method worked and the boiler did not leak.

Smiles's narrative has been repeated elsewhere, and while it all sounds plausible, it does not agree with Phipps's recollection:

> All I can say is that I believe this [Smiles's account] to be quite erroneous, and that such a method was never tried.
>
> There was much discussion as to the best method for securing the tubes, and quite probably the above was discussed, but it was not actually tried – indeed, to have done so and then changed to the ferrule system would have implied the formation of new boiler ends, which certainly was not done.

The first step in attaching the tubes using the ferrule system was to drill the holes in the tube-plates to the same size as the outside diameter of the tubes. The tubes were then inserted into the boiler barrel, probably by feeding them one at a time through one of the tube-plates. Each tube was adjusted until its end was at least flush with the outside surface of the tube-plate. The final stage was to insert a ferrule – a slightly tapered steel collar, something like a thimble with the top removed – into each end. The ferrules were sized to fit inside the copper tubes so that their widest part was greater than the tube's internal diameter – something like a cork being pushed part-way into the neck of an opened bottle. Copper is soft and malleable, much more compliant than steel, so when the ferrules were hammered home into the ends of the tubes, the latter were flared, jamming them tightly against the hole in the tube-plate to give a steamtight seal.

Robert Stephenson kept Henry Booth well informed of progress as *Rocket* neared completion with a series of letters. Writing on 3 August 1829 he told him:

> arrangements have been made which I expect will enable us to have the premium Eng[ine] working in the Factory – say this day 3 weeks – this will give us time to make experiments or any alterations that may suggest themselves. The tubes are nearly all made, the whole number will be completed by tomorrow night, they are an excellent job – the only point I consider at all doubtful is the clinking [securing the ends of the tubes into the holes in the tube-plates] of the ends of the tubes. The body of the boiler is finished and is a good piece of workmanship. The cylinder and other parts of the Engine are in a forward state. After weighing such parts as are in progress the following is an Estimate of the weight . . .

He then listed the weights of the individual components, arriving at a sum total of four tons.

From data available at the time he could calculate the amount of force required to haul a given load, and since he knew his locomotive would weigh about four tons he knew that load would be about twelve tons. From this he could calculate the piston size needed to generate the necessary force, knowing that the operating pressure would be 50 psi. By referring to data for other engines, he would be able to estimate the power output for each piston. This would allow him to calculate the required wheel diameter and piston stroke to achieve the minimum speed of 10 mph. However, to be on the safe side, he based his calculations on a slightly higher speed, and underestimated the locomotive's efficiency. His estimate of boiler size would also have been based on available information, but this was for flue-tube boilers and he had no way of knowing how the new multi-tubular boiler would compare. As it happens the steaming capacity would be considerably better, so the boiler he designed was well in excess of requirements.

Having calculated the forces that would be acting between the driving wheels and the rails, he decided to distribute the weight so that they would carry two and a half tons – just over half the total – to provide sufficient traction. This was achieved by placing the carrying wheels further back, throwing more weight on to the front wheels. He was anxious to know whether Booth thought there could 'be any fatal objection to this'. He also asked him to have the tender built by a particular coachbuilder in Liverpool, as they would 'make one neater than our men'. He went on to say they were 'daily expecting the arrival of the fire box'.

The firebox Stephenson designed for *Rocket* was another unique feature. Conventional locomotives used one end of the flue tube as the furnace, but in the absence of a flue the furnace had to be a separate entity, attached to the back of the boiler. And, rather than being a mere container for the fire, it was designed as an extension of the boiler. The firebox was attached to the rear tube-plate of the boiler and stepped down in level, so that the hot part of the furnace, just above the level of the burning coke, was on a level with the tubes. The hot gases would therefore be drawn straight through into the boiler tubes by the draught set up in the chimney. The front plate of the firebox was accordingly designed to attach to the rear tube-plate without covering the tube openings.

The floor of the firebox was simply an iron grate, with the bars and crossbars spaced sufficiently close to contain the lumps of coke but far enough apart to allow the ash and cinders to fall through. The grating provided a good flow-through of air to keep the fire burning brightly, provided it was kept clear of clinker. Stephenson knew that only a portion of the heat from the furnace would be drawn through the boiler tubes – the rest would be lost through the roof, sides and back of the firebox, even if these were insulated with firebrick. He wanted to wring every last scrap of heat from the furnace, and the best way of doing so was to surround it with a water jacket. That way the heat would be used to raise the temperature of the water, rather than being lost to the atmosphere. The best material to use for the water jacket was copper because of its superior conductivity. Ideally the sides, roof and back of the firebox would be constructed from double-skinned copper sheeting: there was no need to do anything about the front because that was in contact with the boiler. The water inside the water jacket would be in continuity with that in the boiler.

If there had been enough time he would have designed the ideal firebox, but expediency called for compromise. In consultation with his father he designed the sides and roof as a single unit, aptly named the saddle, constructed of ⅜ in. and ¼ in. thick copper plate. Stays were used to join the outer and inner skins together, and these were set at 3-in. intervals, which was also the thickness of the water jacket. The back of the firebox was a plain wrought-iron plate fitted with a hinged door for stoking the fire. The contract for building the saddle was awarded to Liverpool coppersmiths, Messrs James Leishman and John Welsh, who completed the job at a cost of £58 2*s* (£3,400).

A pair of copper pipes, one on either side, connected the bottom of the water jacket with the lower half of the boiler. A second pair of pipes connected the top of the water jacket to the top half of the boiler. When the locomotive was in steam there was a continuous circulation of water between the two, maintained by convection currents, and the pressure inside the saddle was accordingly the same as that inside the boiler.

In his 3 August letter to Booth, Stephenson mentioned that *Locomotion*'s tyres had recently failed on the Stockton & Darlington Railway. He attributed this to the combined effects of the uneven wear between the front and back wheels and their being joined together by the coupling rods. As the most worn wheels were slightly smaller than the others, they tended to rotate more rapidly, but were forced to turn at the same rate as the others by their coupling rods. Stephenson thought this caused 'a considerable loss of power', a conclusion that has since been confirmed experimentally. *Rocket* would not suffer the same defect because the wheels would not be coupled. Having two-wheel rather than four-wheel drive would reduce *Rocket*'s traction if the track was wet, but that was offset by the increased power.

His letter ended with two postscripts:

> I will write you in a few days detailing Hackworth's plan of boiler, it is ingenious, but it will not destroy the smoke with coal which I understand is intended to form a fraction of his fuel; Coke will be the remainder he does not appear to understand that a coke fire will only burn briskly where the escape of the carbonic acid gas [carbon dioxide] is immediate.
>
> If the two large [driving] wheels having 2½ Tuns [*sic*] upon them is an objection, please inform me, some reduction may perhaps be made, but it must be very little or the friction upon the rail will be inadequate to the load assigned.

The first postscript shows that Stephenson was keenly interested in Hackworth's plans. Hackworth was a talented mechanic with considerable locomotive experience. Having built the highly successful *Royal George*, he was seen as a formidable contender. Stephenson thought Hackworth was planning to burn some coal along with the coke, but all competitors were restricted to coke because of the stipulation in the Railway Act that locomotives must be smokeless. As mentioned before, coke burns less readily than coal, requiring a brisk air supply, and Stephenson obviously thought Hackworth's boiler would be ineffective in this regard.

Stephenson's second postscript underscores his concern that he might exceed an acceptable wheel loading with *Rocket*'s uneven weight distribution. Although the company did not stipulate wheel loadings, it did make it clear that only locomotives weighing four and a half tons or less were permitted to be carried on four wheels. This requirement was to avoid high wheel loadings, and Stephenson wanted to make absolutely sure he would not be jeopardising the entry. Booth, as a senior officer of the company, would be expected to know if this did pose a problem.

Stephenson did not write to Booth again for almost three weeks, after which time there was much progress to report:

> The tubes are all clunk into the Boiler which is placed on the frame. Wheels, springs and axle carriages are all finished. The clinking of the tubes is tight with boiling water. I am arranging the hydraulic permit to prove the Boiler up to 160 lb before proceeding any further. The cylinders and working gear is very nearly finished . . . The fire box is put into its place, but it is not quite square built which gives rise to a little apparent neglect in the workmanship. I have endeavoured to hide it as much as possible. Tomorrow week I expect we shall be ready for trial in the evening.

The reason he wanted to test the boiler at such a high-pressure was because of a stipulation that gave the company the right to test boilers up to a pressure of 150 psi. He had to obtain a permit for the test, though the danger was essentially eliminated by pressurising the boiler with water. Pressurising a boiler with air (or steam) would have been exceedingly dangerous because of the large amounts of energy that would be stored in the compressed gas. Water, in contrast, is incompressible, like other liquids, and does not store any energy, which helps explain why air-filled balloons burst with a bang, while those filled with water only make a mess.

Robert said he hoped to see Booth at *Rocket*'s trial, so they could discuss any alterations that might be required. He hoped his father might attend too, but knew he found it difficult to get away from his office. George Stephenson's time was consumed with completing the railway, as it had been for some long time, which is why he had precious little time for locomotives. Most of the rest of the letter was taken up with a description of Hackworth's boiler, accompanied by a rough sketch. The trouble he took in obtaining weights and measurements of the boiler leaves little doubt that he saw Hackworth as a serious contender for the prize. His familiarity with Hackworth's design was partly because Shildon was only twenty-five miles from Newcastle and because Hackworth's cylinders were being manufactured in the Stephensons' factory.

The hydraulic testing of *Rocket*'s boiler did not go according to plan and had to be halted before reaching half the intended pressure. This was due to an unforeseen construction defect, as he explained to Booth:

> On Wednesday I had the boiler filled with water and put up to the pressure of 70 lb per sq. inch. When I found

> that the yielding of the boiler end injured the clinking of the tubes I therefore thought it prudent to stop this experiment until we get some stays put into the boiler longitudinally. The boiler end at 70 lb per sq. inch came out full 3/16 of an inch. This you may easily conceive put a serious strain on the clinking at the tube ends. Today I had the pressure up to a little above 70. The tubes were nearly every one tight, but the deflection of the end still was more than it was prudent to pass over I am therefore putting in 5 more stays which I believe will be effectual.

This brought the number of stays to at least ten. They ran the entire length of the lower half of the boiler, functioning like extra long iron bolts to tie the two tube-plates together. They were fitted with turnbuckles, which could be tightened when the boiler was cold, stressing the stays and reducing their expansion when hot. The modification was successful and the boiler no longer bulged sufficiently to cause leaking. Although the boiler was now more rigid Robert was reluctant to exceed a pressure of 100 psi.

Industrial espionage, while not common, was nothing new in Stephenson's time. During the Napoleonic Wars several spies were apprehended with information about Britain's manufacturing processes. But it was peacetime, and the man caught snooping in the Forth Street Works was one of the other Rainhill competitors. The incident took place on the morning of 31 August 1829, as Stephenson recounted to Booth:

> Mr Burstall Junior from Edinburgh is in N[ew]castle. I have little doubt for the purpose of getting information. I was extremely mystified to find that he walked into the

manufactory this morning and examined the Engine [*Rocket*] with all the coolness imaginable before we discovered who he was. He has however, scarcely time to take advantage of any hints he might Catch during his transient visit. It would have been as well if he had not seen any thing.

In spite of his previous misgivings about high-pressure testing, Stephenson raised the hydraulic pressure to 120 psi. The boiler performed well, but he still added two more stays, hoping it would then be able to withstand a pressure of 150 psi – Robert Stephenson was cautious by nature and did not take chances. The mercury pressure gauge was almost finished, and the wheels were painted yellow, 'in the same manner as coach wheels . . .'. They looked 'extremely well', and he intended using the same livery throughout. This would make the locomotive 'look light which is one object we ought to aim at'. Yellow, favoured by the fastest stagecoaches, was deliberately chosen to associate *Rocket* with speed.

Booth was unable to attend *Rocket*'s trial run, but received a full report, written in Robert's familiar sloping hand:

Newcastle [upon] Tyne Sept. 5th 1829

Dear Sir

I dare say you are getting anxious but I have delayed writing you until I tried the engine on Killingworth Railway. It appeared prudent to make an actual trial and make any alterations that might present themselves during an experiment of that kind. The fire burns admirably and abundance of steam is raised when the fire is carefully attended to. This is an essential point because a coke fire when let down is bad to get up again . . . We also found that from the construction of the working gear that the Engine did not work so well in one direction as in the other, this will be remedied – The mercurial guage [*sic*] was not on, not from any defect but from my wish to get the Engine tried. We started from Killingworth Pit, with 5 waggons weighing 4 Tuns Add to this the tender and 40 Men we proceeded up an ascent . . . at 8 Miles per hour after we had fairly gained our speed . . . on a level part laid with Malleable Iron Rail, we attained a speed of 12 Miles per hour and without thinking that I deceived myself (I tried to avoid this), I believe the steam did not sink on this part – On the whole the Engine is capable of doing as much if not more than set forth in the stipulations – After a great deal of trouble and anxiety we have got the tubes perfectly tight . . . On Friday next [11 September] the Engine will leave by way of Carlisle and will arrive in L[iver]pool on Wednesday week [16 September].

I am Dear Sir,
Yours faithfully,
Rob Stephenson

	Tn cw qr
The weight of the Engine complete	3. 10. 1
water say	15. 0
Tuns	4. 5. 1

When *Rocket* arrived back at the factory after her trial run some experiments were conducted on the paired blast pipes which discharged exhaust steam into the base of the chimney. By connecting one end of a tube to this region, and the other to a glass tube placed in a bucket of water, they could measure the degree of vacuum formed inside the chimney by seeing how far up the glass tube the water was drawn. Constricting the openings of the copper blast pipes by hammering their ends together increased the force of the blast, creating more of a vacuum. And so, by trial and error, they arrived at the desired force. *Rocket* was now ready for Rainhill, but had to be completely dismantled for shipment.

Two locomotives, built for the Baltimore & Ohio Railroad, and a stationary engine for the Liverpool & Manchester Railway, were finished at the same time. It was planned for *Rocket* to be loaded aboard their vessel, for shipment to Liverpool docks, but her departure was delayed. This was most fortunate because the ship was lost on the treacherous passage around the north coast of Scotland.

When *Rocket* was uncrated after its arrival at the railway company's Millfield Yard, it was discovered that her back wheels had been damaged. A pair of replacements arrived, the day before the trials, but that was not the end of the problem:

> When these wheels were placed under the frame it was found . . . that the journals were too large to fit the bearings. It was impossible to get them altered in time for the trial; but the difficulty was overcome by substituting a pair of cast iron wheels, with square-ended axles . . . taken from a tip wagon at the last moment; and on these she ran at the Rainhill trials . . .

Messrs Rastrick and Wood stoically manned their posts for over three hours while *Rocket* steamed back and forth to complete her first journey of thirty-five miles. She crossed the finishing line somewhat before 2.00 p.m., by which time the other officials had probably already enjoyed their lunch. Very likely the stoics had been served some refreshment during their long ordeal, but if they had wanted to relieve themselves they would have been forced to wait, then make a frantic dash during the fourteen minutes and thirty-four seconds it took *Rocket*'s crew to take on fresh supplies and be on their way again.

Her average speed for the first thirty-mile 'fast' segment was only a little over 13 mph. But her lap speeds steadily improved with her crew's confidence and familiarity with handling her, and she bettered 20 mph on the penultimate leg. The improvement continued during the second segment, averaging about 1 mph faster and reaching over 24 mph on her fastest leg, which was again the penultimate one. She was consistently faster, by about 3 mph, travelling forward, than in reverse. The problem was caused by a small error in the setting of the valve events, that had not been properly fixed after Killingworth.

Rocket finished her trials just after five o'clock, having completely satisfied the judges' requirements. At the end of the day's business Timothy Hackworth approached the judges with a request for more time to work on *Sans Pareil*, declaring he could not get his locomotive ready that week. The judges listened sympathetically. They thought it unlikely *Novelty* would be ready before the weekend either, so they decided to postpone the trials until the following Monday. Braithwaite was not present, so the judges 'agreed with a Friend of Mr Braithwaite's . . . that they should enter upon the trial of their Engine on Monday morning'.

CHAPTER 8

The people's choice

Borrowing a term from the legal profession, the newspapers described Friday 9 October, the fourth day of the trials, as a *dies non*, a non-day, a day when the law courts do not sit. Some spectators had already made their plans to attend and came to the grounds anyway. The most distinguished among these were the Marquis and Marchioness of Clanricard, who later that day would be sailing from Liverpool to Dublin aboard the steam packet *Manchester*. The Irish nobleman was shown around the grounds by one of the officials, but there was nothing much to see, aside from the repairs being conducted upon three of the four locomotives in the competition.

Timothy Hackworth worked diligently on *Sans Pareil*, knowing he had only two or possibly three days in which to make her ready for the trial. *Novelty* was due to compete on the following Monday, so in all probability he would not be called upon until the day after that. But if something untoward should happen to Braithwaite and Ericsson's engine on their appointed day he might have to step into the breach, so he ought to be ready for the Monday. That left only two days: working on the Lord's Day was quite out of the question.

The leaking boiler was his first priority, but after that there were all the other jobs to do that had not been taken care of in Shildon. Most of the necessary adjustments could be made only after steam had been raised and *Sans Pareil* had completed her test run. Running her on the track might reveal yet new problems, but that was beyond his control: his future lay in the hands of the Lord.

Timothy Burstall was not forthright and sincere like Hackworth and his intelligence mission to the Forth Street Works had shown him to be a downright scoundrel. With his own locomotive temporarily out of commission he had been able to get some measure of the competition without revealing his own hand. Having seen the sterling achievements of *Rocket*, and some of *Novelty*'s cut and dash, he was painfully aware of the limitations of his own entry. But what could be done about it?

Ever since the start of the trials he had been busily working away on his locomotive. To all intents and purposes he had been repairing the damage *Perseverance* had sustained in transit from Millfield Yard, with the full approbation and sympathy of the judges. But it seems he had been using this time to alter his locomotive in an attempt to improve her performance. In essence this gave him an extra week to prepare his engine for the trials, with the opportunity of incorporating any new features seen during his observations of the other engines.

Messrs Braithwaite and Ericsson's problem was to complete the repairs to *Novelty*'s damaged 'air-compressing apparatus'. The extent of the damage is not known because no account was ever given. Nor were there any contemporary descriptions of the apparatus itself. According to the *Mechanics' Magazine* it was a 'bellows-sort of apparatus'. The air pump might have remained a mystery but for a chain of events set in motion by a BBC television programme. The *Timewatch* series was staging a re-enactment of the Rainhill trials, scheduled for the autumn of 2002, between the working replicas of *Novelty*, *Rocket* and *Sans Pareil. Novelty*, on loan from the Swedish Railway Museum, had been shipped from Sweden for the occasion, but because she had not been in steam for many years an overhaul was needed, including inspection and certification of the boiler. This was carried out at the National Railway Museum in York, which, apart from having one of the most spectacular collections of locomotives in the world, has a well equipped workshop where major repairs can be undertaken. When the locomotive had been restored to working order it was sent to Llangollen to join the others.

Contemporary illustrations of the original locomotive depict the bellows apparatus as being housed in a large copper box, shaped like a hopper, which extended across the full width of the locomotive, at the opposite end to the boiler. When the replica was built in 1979 it was assumed that the distinctively shaped box housed a large flap that flipped back and forth, sucking in air on one stroke and blowing it out on the other, something like the action of a pair of bellows. The *Novelty* replica was accordingly built with such an apparatus, driven mechanically through a linkage to the right-hand piston. The disadvantage of this apparatus is that the air flow through the boiler is pulsed rather than continuous, which is not ideal for steam production.

One of the technically minded observers at the Llangollen trials was interested in the history of lead smelting, and about two weeks after the trials he happened upon a patent specification for an air-blowing apparatus, invented by a man named Vaughan. The singular feature of Vaughan's patent air blower was that it gave a continuous air flow, making it especially suitable for lead smelting, and other applications where a pulsed flow is undesirable. The drawing depicted a device *exactly* like the one shown in contemporary illustrations of *Novelty*. The timing was right too because Vaughan's invention pre-dated Rainhill by a couple of dozen years. Here was the solution to a 175-year-old mystery. Full details of this remarkable discovery will be given at the third International Early Railway Conference, in the autumn of 2004. Suffice it to say here that the hopper-shaped chamber was divided in two, with two separate impellers. These pumped out of phase, so as one side of the apparatus was sucking the other side was blowing, giving a continuous air flow, like two pairs of bellows being used. The left side would have been driven by *Novelty*'s left-hand piston, the right by the right-hand piston. The original *Novelty* would therefore have performed better than her replica, simply because the air blower had been wrongly reconstructed in 1979.

While Messrs Hackworth, Burstall, Braithwaite and Ericsson worked away on their respective engines, George Stephenson probably used the day trying to catch up on his regular duties. He had the full responsibility of completing the Liverpool & Manchester Railway on time, and with so much still left to do he could ill afford to take time away from the job. A pragmatist like Stephenson probably resented the time spent at Rainhill, but it was a pleasant and undemanding diversion from his normally hectic life.

The Rainhill trials were demanding on the judges as well as the contestants and Friday, the *dies non*, must have been a welcomed break. Not that they indulged themselves in a rest because they repaired to Liverpool that morning for a judges' meeting. They were not expecting to see any of the competitors and were therefore surprised at an unexpected visit from Mr Braithwaite, 'in company with his friend'. The friend was not identified but may have been Charles Vignoles; Ericsson was probably still at Rainhill working on *Novelty*. Braithwaite 'declared that his Engine would be all complete and perfectly ready for entering upon the task on Saturday morning, and insisted that his Engine should be put upon trial on that Day . . .'. But the judges had their misgivings. Rastrick and Wood, both experienced locomotive engineers themselves, were well aware of the work that had to be done to complete *Novelty*'s repairs. 'Several joints were to be made, which it would be almost impossible to get done in time to allow of their Setting . . .' A postponement made all the sense in the world, and the judges recorded in their report that 'we did every thing in our Power to defer the trial until the Monday . . .'. But their efforts were to no avail, so they acquiesced to Braithwaite's wishes. *Novelty* would be put to the test first thing on Saturday morning: the very next day.

Large crowds gathered to see *Novelty*, their favourite, undergoing her trials, and there was every expectation of her bettering *Rocket*'s performance. It was a cloudy day, but at least it was mild, with temperatures that would reach 56° F by midday, which was almost ten degrees higher than it had been so far. The large attendance was partly attributable to the weekend, though most working people did not have free Saturday mornings. The enthusiastic spectators were ready, the judges were ready, but *Novelty*, it appeared, was not. However, after some delay the boiler was filled with water and the locomotive was wheeled on to the scales.

Deducting the weight of her water tank, and its contents, neatly tucked away beneath her platform, she weighed in at 3 tons 1 cwt.

Now it was the judges' turn to be unready, and they went into a lengthy discussion of what was the fairest way of assigning an appropriate load to the unconventional locomotive. In the end they decided to base her burden on the load that *Rocket* had hauled, pro rata according to the differences in their respective engine weights. They assigned *Novelty* a load of 6 tons 17 hundredweight and two carriages loaded with stones were hitched to the locomotive.

Steam was raised in fifty-four minutes and forty seconds. This was barely two minutes faster than *Rocket*, but the London engine had consumed a little less than half as much coke in the process. As the expectant crowd looked on, *Novelty*, and her assigned load, were manhandled to the starting point.

The signal was given and *Novelty* started away from the line at precisely twenty-eight seconds past eleven o'clock. Gathering speed, she reached the first post one minute and twenty seconds later, beating *Rocket*'s best start by five seconds. The crowd was ecstatic and showed its appreciation with enthusiastic applause.

Novelty was moving much faster than *Rocket* had done at the start of her ordeal, and if any gentleman had been timing her progress with his pocket watch he would have seen that she completed her first dash between the two timers' posts in a little more than five minutes. The official speed on this first leg at just over 16 mph, almost on a par with *Rocket*'s third best time of her entire first thirty-five-mile journey. Her second leg was somewhat slower, but still exceeded 13 mph and was better than half of *Rocket*'s first set of runs. This was a most promising start and *Novelty* seemed all set to snatch the prize away from the Stephensons. But she had barely passed the first post on the commencement of her third leg when calamity struck. The pipe taking water from the feed pump to the boiler suddenly burst, sending water 'flying about in all directions'. According to the entry Rastrick made in his notebook, the mishap was 'in consequence of the Engine man having shut the cock between the forcing Pump and the Boiler'. Fortunately the water was quite cold – had it been scalding water from the boiler Messrs Braithwaite and Ericsson might have been grievously injured. *Novelty* coasted to a halt and a brief meeting was convened with the judges.

The extent of the damage was only minor and the pipe could easily be repaired, but there were no facilities for doing the job on the grounds. The pipe therefore had to be sent to the village of Prescott, about two miles away, where there was a workshop. The *Mechanics' Magazine* was critical of these organisational shortcomings, and expressed the hope that the officials of any future competition would provide a forge for making repairs on the spot, fully equipped with all the necessary tools and materials. Meanwhile *Rocket* entertained the disappointed crowd with an impressive demonstration, as it had done the last time *Novelty* had broken down.

There is no record of who was driving *Rocket*, but the demonstration was marked by a certain flamboyance that might have been George Stephenson's hand. With its tender uncoupled, the locomotive was twice run back and forth across the measured section of the track, seven miles in all, to show what it was capable of achieving when unencumbered. The reporter from *The Times* allowed himself to be quite carried away by his account of the display:

> the engine alone shot along the road at the incredible rate of 32 miles in the hour. So astonishing was the celerity with which the engine, without its apparatus, darted past the spectators, that it could be compared to nothing but the rapidity with which the swallow darts through the air. Their astonishment was complete, every one exclaiming involuntarily, 'The power of steam is unlimited!'

Travelling at such a speed in *Rocket* was an alarming experience, as confirmed by the crew of the replica engine at the Llangollen re-enactment. The locomotive bucked and jumped so much that it was difficult to hang on, and the noise and the rushing of the air made it seem as if it was travelling considerably faster.

By the time the broken pipe had been repaired and refitted it was too late for *Novelty* to undergo her trial, but the judges assured Messrs Braithwaite and Ericsson that they would be given another chance to compete. As the afternoon was not yet done *Novelty* was put through her paces for the gratification of her ardent supporters, which seemed to include most of the crowd. This time Charles Vignoles climbed aboard to act as timekeeper. Hauling her assigned load of stones, the locomotive made a return trip across the measured section of the track. Vignoles gave a detailed report of her times as she passed the various landmarks along the way, from the starting post to the turning post, via the judges' tent and the grandstand. On her fastest run between the two timekeepers' posts he recorded a speed of 17½ mph. A carriage with seats was then substituted for the wagons loaded with stones, and, as he reported in an article that appeared both in the daily newspaper, *The Albion*, and in the weekly *Mechanics' Magazine*:

> about forty-five ladies and gentlemen ascended to enjoy the great novelty of a ride by steam. We can say for ourselves that we never enjoyed any thing in the way of travelling more. We flew along at the rate of a mile and a half in three minutes; and though the velocity [30 mph] was such, that we could scarcely distinguish objects as we passed by them, the motion was so steady and equable, that we could manage not only to read, but write.

Vignoles was not the only man aboard the improvised train with an interest in timing its performance. The reporter for the *The Liverpool Chronicle* recognised Dr Traill and his family among the passengers:

> The doctor timed the speed of the *Novelty* while running the full course, and it appears to have averaged twenty-two miles an hour, with forty-five passengers, and at one period carried the same passengers at the inconceivable velocity of thirty-two miles an hour. [At the end of the experiment] the company hastened back to town, to await the races of this week, with an impatience only to be conceived by those who saw with their own eyes what locomotive engines can do.

Novelty's remarkable performance that afternoon must have attracted more attention than that of the curious spectators, and it is difficult to imagine how any of the other contestants could have witnessed the event impassively. Perhaps George Stephenson would have stuck doggedly to his earlier dismissal of *Novelty* on the grounds of her lack of 'goots'. But Robert, who had not inherited his father's bold confidence, may have felt threatened. The pious Hackworth may have felt as much admiration as misgiving for their stellar performance – when Messrs Braithwaite and Ericsson had first arrived at Rainhill, the devout Methodist had apparently given freely of his considerable mechanical skills to help them ready their locomotive for the trials. Burstall's thoughts, most assuredly, would have been less benign, and he may have wished for another calamity to befall the infernal machine.

Sans Pareil was next on the judges' list. After consulting her builder, still busy with his preparations, they decided that his trial would take place on Tuesday morning. Hackworth stoically assured them that 'if it did not succeed on that day he would not require any further time'. There would be no activities on the morrow, it being the Sabbath, and the Monday following was declared a second *dies non.*

CHAPTER 9

The dark horse

Rainhill without the crowds. Rainhill without all the cheering and the band-playing and the noise. No frock-coated gentlemen, no crinolined ladies, no labourers in their Sunday best. Not a single juggler or card-sharp in sight. No activities had been planned for the day and none of the judges or directors appeared on the grounds. Nonetheless, a few spectators had shown up that Monday, hoping 'that some novelty might be witnessed'. However, aside from some of the engines being 'exercised', there had been little to see, and the last stragglers probably drifted away before evening began to fall. Timothy Hackworth probably had it all to himself. But he had no more time to enjoy the peace and tranquillity of the evenings than he had to take in the hurly-burly of the trials during the days. The Lord's Day notwithstanding, he had been hard at work since he arrived.

Mercifully the boiler was no longer leaking as it had been, and *Sans Pareil* would be ready to face her fate in the morning. The evening's test run had gone very well and he was confident the sturdy locomotive would give a good account of herself, more than satisfying the judges' stringent tests.

Novelty may have been light and swift, but would her small boiler give her the necessary staying power to complete two thirty-five-mile journeys, back-to-back? Hackworth was fairly familiar with her design after helping Braithwaite and Ericsson get her ready to compete, and was probably of the same opinion as George Stephenson regarding her lack of 'guts'. As for the Stephensons' own entry, Hackworth was convinced that, in spite of the considerable advantages they enjoyed, his was still the better machine.

Her simple construction was founded upon the proven principles he had used in designing his much celebrated *Royal George*. Indeed, *Sans Pareil* was very much a shortened version of that powerful engine, built on four wheels rather than six. Her 6 ft-long boiler was just under half that of *Royal George*, but its diameter was only 2 in. less, giving the locomotive its squat appearance. As in other flue-tube boilers, the furnace occupied the end of the flue. However, whereas in previous locomotives, including *Royal George*, the flue tube ended flush with the boiler barrel, Hackworth extended it by some three feet, to increase the heating area. The boiler casing was extended around the top and sides of this projection forming a half-shell capping. This capping formed a water-filled continuation of the boiler barrel, designed to extract more heat from the fire. While it made some contribution to heating the water in the boiler, it was far less effective than the copper saddle that encased *Rocket*'s firebox. *Sans Pareil* had a similar extension around the base of her chimney, as shown in the sketch John Rastrick penned in his notebook. The grate, comprising iron bars, was positioned at about the mid-point of the 2 ft-wide flue tube. The upper half of the end of the flue tube was capped by a semicircular cover, probably fitted with a small door for stoking. The gap beneath the grate formed an air passage for the fire.

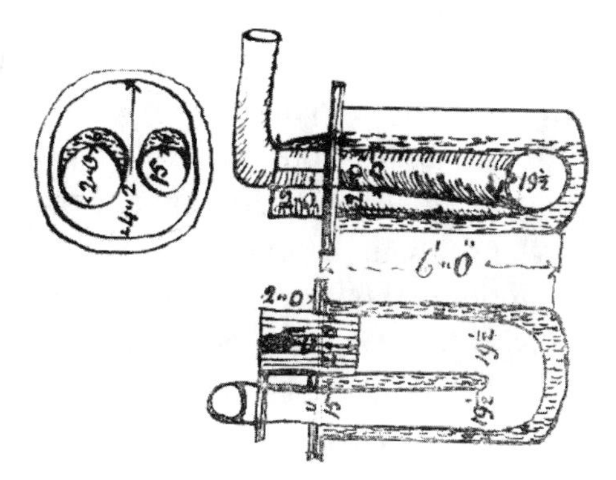

Rastrick's sketch of Sans Pareil's boiler, showing the extension of its casing around the top and sides of the furnace and chimney base.

San Pareil's cylinders were not as large as those of her forerunner, and were only 7 in. in diameter compared with 11 in. in *Royal George*. This gave them only about 40 per cent of the piston area, so they could generate just under half as much force. But *Sans Pareil* was a much smaller engine and Hackworth was sure the pistons would be more than adequate for the job.

Hackworth may have tarried for a while beside his engine that night, listening to the metallic clicks and creaks as the hot flue tube cooled down. She might have been a living creature, a beetling iron creation all of his very own, conceived, designed and paid for. And tomorrow, God willing, she would deliver a winning performance.

The judges arrived on the grounds soon after eight o'clock the following morning. It was a wet and dreary start to what promised to be a cheerless autumn day, but at least it was fairly mild. Hackworth had been there for some time, making his last-minute checks. He was soon joined by Messrs Rastrick, Wood and Kennedy. They fully expected to find the locomotive's boiler filled with cold water in readiness for the weighing-in, as set out in their printed rules. They were therefore somewhat dismayed to find the boiler still hot from the previous night's run. Emptying the hot water from the boiler and refilling it with cold would have taken too long. So, in the interests of expediency, they decided to forgo timing how long it took to raise steam and how much coke was used in the process. These would have to be determined on another occasion.

With many strong arms and backs to assist, *Sans Pareil* was wheeled on to the company's weighing machine. Being so short, she fitted on the scale with room to spare, so they did not have the same difficulty they had with *Rocket*. But a much worse problem lay ahead. When the judges checked their figures they discovered that *Sans Pareil* was too heavy. According to the stipulations and conditions laid down by the directors, the weight limit for a four-wheel locomotive was four and a half tons. But Hackworth's engine weighed 4 tons 15 cwt 56 lb, which was more than 600 lb over the limit. Hackworth vehemently argued that the weighing machine was incorrect and that his engine was not overweight. But no matter how much he protested the judges remained resolute in their decision that 'under these circumstances it could not be considered as entitled to contend for the Premium . . .' On reflection they decided to let him compete, leaving it to the directors to make the final decision on his eligibility if he proved successful. *Sans Pareil* was assigned a load of 14 tons 6 cwt 56 lb, making a total of just over nineteen tons with the tender, compared with *Rocket*'s seventeen tons. The heavy train was pushed to the starting point, steam was raised and *Sans Pareil* was under starter's orders.

Because of her return-flue boiler, driver and fireman rode at opposite ends of the engine. The fireman, standing in the tender, faced the front of the engine, and had to lean forward across the gap between the two hitched vehicles as he bent down to stoke the furnace. He was alternately pushed backward when the locomotive moved forward, and then drawn forward when the engine was running backward. With nothing between himself and the roaring furnace and searing chimney, his situation could hardly have looked more hazardous. But the driver's position at the back was no less perilous, balanced on a narrow wooden platform 6 ft in the air, poised above the two driving wheels and just inches from the thrusting connecting rods With no safety rail he had to hang on to the valve gear as best he could. If he fell on to the track when the engine was travelling backward the cold iron wheels would slice through his body like a guillotine. But neither man would have given such perils a moment's thought, especially as they waited for the starter's flag to fall. The crew aboard *Sans Pareil* appears to have been Timothy Hackworth, probably assisted by William Gowland, one of his most trusted drivers. The day had not started well for Hackworth, but at least they were now all set to get on with the job. A great deal of time had been lost over the weighing-in problem, but a little after ten minutes past ten o'clock the signal was given and they were away.

For all her ungainly stubbiness *Sans Pareil* got off to a flying start, and it 'was soon manifest that a very powerful competitor had entered the field'. She tore past the first post in one minute and nine seconds. This remarkable performance even beat *Novelty*'s record-breaking start, by a margin of eleven seconds. This was the fastest start of the entire competition, and Mr Rastrick would have been forgiven for taking a second glance at his watch to ensure his eyes did not deceive him. Maintaining the blistering pace, she completed the first leg of her journey at a speed of 17½ mph. This was faster than either of *Novelty*'s laps, and was equal to the second highest speed recorded on either one of *Rocket*'s thirty-five-mile journeys. Her second lap was slower, at just under 12 mph, but she picked up the pace again during the third one, reaching almost 15 mph. And she kept dashing back and forth across the course, like a sheepdog rounding up sheep.

Sans Pareil's speeds, like those of *Rocket*, were consistently faster when travelling forward than in reverse, and she was evidently experiencing a similar disparity in valve events. But it was not adversely affecting her performance, and she was outperforming *Rocket* by a small but significant margin. The audience had so far seen very little of *Sans Pareil* because of the time spent in repairs, and her sudden appearance in the spotlight must have taken many people by surprise.

She was not a graceful performer and John Dixon, the engineer, wrote rather disparagingly that she 'rumbles & Roars and rolls about like an Empty Beer Butt on a rough Pavement . . .'. Her rolling motion was caused by the vertical orientation of her cylinders. As each piston thrust down upon its driving wheel there was an equal thrust in the opposite direction: the reaction. This raised that side of the locomotive and these alternate reactions to the piston's thrust caused the engine to roll from side to side. If the cylinders had been placed horizontally the reaction forces would have attempted to make the locomotive yaw (fishtail). Cars often yaw, especially on slippery roads, but rails resist yawing movements in locomotives. *Rocket*'s inclined cylinders reduced the rolling effect, and, as mentioned earlier, it was later eliminated by repositioning them nearly horizontally.

Sans Pareil's loud roaring from her chimney was due to the powerful force of her blast pipe. Hackworth – sometimes erroneously credited with inventing the blast pipe – was a great believer in the device, and tended to be excessive in its use. As superintendent of locomotives for the Stockton & Darlington Railway he had all the blast pipes modified by constricting their apertures. The force of the blast became so great that sparks and ashes were ejected from the chimneys, which became something of a Hackworth hallmark. Predictably, he used the same strategy in *Sans Pareil*, and 'the draft [*sic*] was so great as to carry with it the fuel from the chimney . . .' Robert Stephenson had similar recollections of *Sans Pareil*, noting that, 'nearly half the fuel was thrown out of the chimney unconsumed, as many can testify who witnessed the experiments at Rain Hill'.

Her wasteful appetite would have kept the fireman hard at work, and his job was already difficult and dangerous enough. Unlike the driver, he had no designated place to stand and had to position himself, as best he could, at the front of the tender, clearing some standing room amidst the coke. The 2 ft gap over which he leaned was open, with no sort of footplate to stand on. If he lost his footing and fell through the gap his chances of being crushed by one of the wheels was exceedingly high. The water jacket capping the projecting flue tube, which helped heat the water, prevented the outside from getting much hotter than scalding water, but, the chimney, towering above his head only a foot or two away, was scorching hot.

The fire grate continued forward beyond the far end of

the projecting flue tube, extending about 2 ft into the boiler barrel. The grate was therefore about 5 ft long and 2 ft wide, which was quite small for the boiler. Such a long narrow grate presented the fireman with a major problem. No matter how far across the gap he leaned, there was no possibility of reaching the far end of the grate with his shovel. But he had to keep the entire grate covered with burning coke to generate sufficient heat. And if there were any gaps the draught through the bottom of the grate would pass right through them, and be effectively wasted. One way of dealing with the problem was to build up a good fire in the middle of the grate, then push the hot coke forward with a fire-iron. Aside from stoking the fire he also had to keep the passageway beneath the grate clear of the ash and cinders falling through the bars, otherwise the draught was stifled, the fire died down and steam pressure fell. He also had to keep the grate clear of clinker, as best he could.

Tending the furnace would have been a full-time job during the rigours of the competition, but the fireman had other duties as well. He was responsible for checking the water barrel at the back of the tender, and, more important, the water level in the boiler. Modern locomotives have a glass gauge, called a sight gauge, on the boiler for the purpose. It is easily read and *Rocket* may have had one at Rainhill, but *Sans Pareil* used the older method of try-cocks. These comprised a pair of taps, set into the boiler one above the other and separated by a few inches to span the optimum water level. The fireman opened the try-cocks sequentially, starting with the top one, and if he got water instead of steam the water level was too high. This was not serious because he could always drain off the excess, if so inclined. However, if he opened the bottom cock and got steam instead of water, he knew the water level was too low. This was a critical warning and he had to replenish the boiler immediately. But *Sans Pareil*'s feed pump was at the other end of the boiler, accessible to the driver, not the fireman, who would activate the pump at the fireman's request. This had to be done with the engine travelling slowly. While inconvenient, this was not difficult: the notion that early locomotives were incredibly noisy is quite erroneous and driver and fireman could easily converse, even at speed. However, reaching the try-cocks in the first place was a challenge because they were positioned on the boiler end-plate. To do this the fireman had to cross the gap between tender and engine and stand on the top of the hot water jacket that capped the flue tube. With nothing to hang on to but the try-cocks themselves, this would have been difficult and dangerous when the train was moving. The Rainhill fireman probably postponed doing it until the engine paused at the end of one of its legs, but if the locomotive were in service and on a long run he would have had no choice. Workplace safety was not an issue in the early 1800s, and workers routinely exposed themselves to risk without a moment's hesitation.

Regardless of her hacking cough and poor deportment *Sans Pareil* was a lusty performer and her outstanding debut would not have been missed by the Stephensons. Robert recognised Hackworth as a serious challenger months before the trials, and his father always had a high regard for Hackworth's abilities, often seeking his opinions on technical matters. Although George had been dismissive of *Novelty*, he may have been disconcerted when he saw *Sans Pareil* settling in for her second hour of relentless steaming.

If George Stephenson had timed her first hour's performance and compared it with that of his own entry, he would have found *Rocket* was 2 mph slower. *Rocket*'s performance had improved during her second hour – supposing *Sans Pareil*'s performance also improved with time? What if she soundly thrashed *Rocket*? Perhaps such thoughts never crossed the mind of the self-assured patriarch. But if the unthinkable *did* happen he stood to lose a considerable amount of money. And so did his partners. The £500 purse paled into insignificance compared with the contract to supply locomotives to the company. Even worse, if *Sans Pareil* triumphed over *Rocket*, Hackworth, not Stephenson, would become the new tsar of the country's fledgling railway system!

And what thoughts might have been going through Hackworth's mind as his indefatigable machine charged back and forth across the course? Perhaps Hackworth was too preoccupied with the business of running the locomotive to spare a thought for anything else. But that seems unlikely. Surely he must have allowed his mind to light upon the subject of winning, at least for a fleeting moment? And if he had entertained such a thought he would surely have considered its implications. Hackworth was not a dreamer like Trevithick and after a brief taste of free enterprise prior to joining the Stockton & Darlington Railway, he appears to have been happy to work for somebody else again. So perhaps a victory at Rainhill would have meant nothing more than a change of job, from running a railway to running a locomotive workshop. The most important aspect of winning the trials for Hackworth would probably have been recouping his personal savings. He was not a man of means like the Stephensons and could ill afford to lose the money he had invested in his locomotive.

Hackworth's immediate problem was the possibility that *Sans Pareil* would be ruled ineligible. The weigh scales must surely have been wrong, but the judges had been adamant about the accuracy of their equipment. What if he should win the trials only to be robbed of his hard-fought victory by the intransigence of the judges?

Regardless of the opposition stacked against him by the Stephensons with all their advantages, his engine was soundly outperforming theirs. Not only were *Sans Pareil*'s lap times during the first hour faster than *Rocket*'s, but her turn-round times were much better too, showing Hackworth's reversing mechanism was as effective as Robert Stephenson's. The judges' records show that some of *Rocket*'s lost time was attributable to her crew. On the first journey they made a brief stop at the end of the second leg to do some oiling, and another at the end of the third lap to grease the pistons that cost them several minutes. They took on six buckets of water at the end of the twelfth leg, stopping for five more buckets three laps later. There were similar delays during her second thirty-five-mile journey, to grease wagons, take on coke and replenish the water. *Sans Pareil*, in contrast, just kept on going, and did not stop until the end of the eleventh lap. But this was an inordinately long pause, adding about twelve minutes to the time as they oiled the wagons and attended to the feed pump. Two laps later they stopped to take on eight gallons of water, and again in another two laps for a similar quantity. This was at the end of her fifteenth lap and it was evident the feed pump was causing Hackworth some concern because he spent some more time working on it. Regardless of the nature of the problem it was not adversely affecting *Sans Pareil*'s performance because this last lap, at a little over 16 mph, was her second fastest.

Sans Pareil thundered past Mr Wood at thirty-seven minutes and thirty-two seconds past noon, heading west towards the grandstand on her sixteenth lap with only four more to go to complete her first thirty-five-mile journey. Hackworth was almost half-way through his trial. But *Sans Pareil* never reached the grandstand. With a sudden explosive whoosh she belched a cloud of ash and steam from her chimney, and coasted to an undignified halt. The locomotive had clearly suffered a devastating failure, but only an engineman among the spectators could have guessed the likely cause of the problem. Hackworth, absorbing the impact like a body blow, knew exactly what had happened.

The chain of events leading up to the calamity began with the failure of the feed pump, which was mechanically driven by an eccentric on the rear axle. As the level in the water-starved boiler fell, it eventually reached the top of the flue tube. Allowing a boiler to run so low in water is dangerous. Once the top of the flue tube loses contact with the water in the boiler, its temperature soars towards that of the combustion gases and burning fuel in the furnace. Heating the metal to such high temperatures causes it to soften and the pressure of the steam above pushes down on the compliant flue tube. In the case of *Sans Pareil* this force was in excess of three tons per square foot. At best this causes the flue tube to bend downward, causing permanent distortion but no failure. But there is every chance that the tensile stresses acting upon the weakened roof as it is stretched downward will cause one or more of the rivets to fail, rupturing the flue tube, sometimes explosively.

As a safety measure the roof of the flue tube was fitted with a fusible plug, a device developed by Trevithick. The plug was made of lead, which has a low melting point, and, as soon as the water level reached the top of the flue tube, it melted. Steam, accompanied by water under pressure, was immediately discharged into the flue tube, releasing the boiler pressure and dousing the fire in the furnace.

Sans Pareil's fusible plug 'melted out near the Grand Stand and the Engine was brought from thence by men to the watering place opposite the Blacksmiths shop where it arrived at 0 hr. 55 min. 0 s [12.55] PM'. Rastrick wrote in his notebook that 'The Plug was put in again and the fire lighted, the cask was filled at 1.30 and the Steam was got up again at 2.15 but the forcing Pump having gone wrong again, the further prosecution of the Experiment was abandoned.' The judges' official report gave a little more detail:

> The forcing pump of the Engine . . . required so long a time to get in order together with the loss of time in putting in a new lead plug and filling the Boiler again and raising Steam that it was too late for the Engine to enter upon its task, so as to accomplish it before dark, and consequently the further prosecution of this experiment was abandoned.

Sans Pareil did not receive as much press coverage as *Rocket* or *Novelty*, but her performance was highly praised. The account in *The Times*, repeated verbatim in *The Albion*, reported that:

> for two hours this engine performed with great speed and regularity, averaging full 14 miles an hour, for a distance of upwards of 25 miles, while drawing that enormous load. Unfortunately one of the pumps which supply the boiler from the tender was out of order . . . This is no proof of failure, because it has nothing to do with the principle; and we are assured, that when The Sans Pareil has got into good working order, she will rank high in the list of competitors, and may well be considered as having fulfilled the original conditions.

Sans Pareil was also mentioned in connection with a potentially fatal incident which took place sometime before the day of her trial: 'a man fell, by some accident, within the rails, when Mr Hackworth's steam-carriage was approaching with great rapidity; but, having had the presence of mind to lie down, the machine passed over him without doing him any injury.' The *Mechanics' Magazine* concluded its coverage of the day's activities with the observation 'We understand the judges subsequently resolved that 'The Sans Pareil' should have another trial on Friday, the 16th.'[15]

According to the judges' report, *Sans Pareil* covered a total distance of 27½ miles before being brought to a standstill, consuming 1,269 lb of coke and 274 imperial gallons of water. By simple proportions they calculated she would have needed 1,615 lb of coke and 348 gallons of water for the entire thirty-five-mile journey. *Rocket*'s consumption, in comparison, was 542½ lb of coke and 289½ gallons of water. *Sans Pareil* therefore used almost three times the fuel and 20 per cent more water than the Stephensons' entry. 'He has to feed her with more Meal and Malt Sprouts than would fatten a Pig', Dixon wrote sardonically. The high fuel consumption was largely due to the excessive force of her blast pipe, which caused unburned fuel to be ejected from the chimney, along with the hot gases from the furnace. Even if the blast had been reduced, fuel consumption would still have been high compared with *Rocket*'s because the boiler was less efficient. There were several reasons for this. Foremost, the heat transfer from the flue tube to the water was far less efficient than it was in the multi-tubular boiler, so much more heat was lost through the chimney. The primary reason was that the surface area available for heat exchange was much smaller in *Sans Pareil*; it has been given as 74.6 sq ft – probably an overestimate – compared with 117.8 sq. ft for *Rocket*.[17] The heat-exchange area within the boiler was therefore almost 60 per cent higher in *Rocket*. The greater thermal conductance of copper over iron also enhanced heat transfer. Moreover, having numerous hot pipes running through the boiler set up strong convection currents in the water, causing good mixing, and this avoided the stagnation of cold water at the bottom, which was an inherent problem of flue-tube boilers. Furthermore, *Rocket*'s copper saddle was a major source of heat transfer, whereas the water jacket capping *Sans Pareil*'s flue tube was far less effective.

High fuel consumption was the price Hackworth had to pay to wring sufficient steam from his boiler to sustain the relatively high speeds needed to win at Rainhill. His locomotive probably represented the upper limit to which a return-flue boiler could be pushed.

With the forcing pump repaired and a new fusible plug in place, *Sans Pareil* was ready for another attempt at the prize. Hackworth was not sure when that might be, but the following morning had been set aside for *Novelty*'s third attempt at the prize.

CHAPTER 10

Winning day

Not since the trials began, eight days earlier, had such a large crowd – estimated at between ten and fifteen thousand – gathered on the grounds to watch the locomotives compete for the grand prize. In contrast to the previous days, Wednesday 14 October was bright and dry and it promised to be a good day. But even if it had been wet and cold the crowds would still have converged on the grounds because it was the day everyone expected the premium to be won. Popular support seemed to be behind *Novelty*, last seen travelling at the unbelievable speed of 30 mph, hauling upward of fifty ecstatic passengers. The mishap over the burst feed pipe had robbed her of the prize on that day, but nothing would stand in her way now.

The earliest spectators had already taken their places in the grandstand or were milling around the grounds when the judges arrived. Others were still streaming in, and would continue to do so for some time to come. The judges expected to find *Novelty* all ready and waiting with her boiler filled with cold water, all set for lighting the furnace and raising steam. They were therefore surprised to find her boiler in several pieces, with workmen busily trying to reassemble it. Mr Braithwaite was not among those gathered around the engine, but his partner was, and if he had some explanation to offer for the unexpected turn of events, the judges did not include this in their report. According to the *Mechanics' Magazine*, the failure of the feed pipe four days before had allowed the water level in the boiler to fall below that of the flanged joint between the furnace and the flue. This had damaged the joint, necessitating dismantling the boiler to repair it. Presumably the damage had not been discovered until after *Novelty* had put on her exhibition for the crowds, following the repair of the burst pipe.

While the men worked away on *Novelty*'s boiler the spectators were treated to a demonstration of *Rocket*'s remarkable power. At each end of the two-mile level section of track where the trials were being conducted was an incline. That on the west, closest to Liverpool, was the Whiston incline, and that to the east, towards Manchester, was the Sutton, each named for a nearby village. The inclines were about one and a half miles long with a gradient of 1 in 96, or about ⅜ in. per yard. Because of the low friction between iron wheels and rails, locomotives experienced problems with gradients, even gentle ones like these. The intention had always been to install stationary engines at the top of the two inclines for hauling the trains to the top. But soon after *Rocket* had arrived at Rainhill, George Stephenson had tried her out and found she coped perfectly well with the gradients. To demonstrate the point publicly he hitched a carriage to the engine, filled it with twenty-five expectant passengers, and set off down Whiston incline at a clip. Reaching the bottom, he reversed the valve sequence and charged back up to the top with great ease, to the sheer delight of everyone aboard. He repeated the performance several more times, leaving no doubt in anyone's mind that stationary engines were not needed on the Liverpool & Manchester Railway and were now obsolete.

It took the workmen the rest of the morning to reassemble *Novelty*'s boiler. Because of the long delay the judges decided to forgo assessing the time and amount of coke needed to raise steam – information already available from the previous trial.

Almost as if the locomotive were anxious to get under way, steam was 'got up . . . in somewhat less than 40 minutes . . .'. She then took a short run up and down the line, just to make sure everything was in working order, before lining up at the starting point. The two carriages with her assigned load were attached, the hitching chain was checked, and the crew gave the nod that they were ready. Moments later the starter's hand fell. They were off. The time was twenty-five minutes and forty seconds past one o'clock.

Novelty got off to a slow start, completing the eastward leg of her first trip at a modest 8 mph. No doubt her driver was keeping her back, to make sure the boiler's joints were holding up. But she doubled her speed on the westward dash. Her adoring crowd was delighted – this was the day they would see their *beau idéal* win the Rainhill trials. The third leg was fractionally slower, at just under 15 mph, but most of *Rocket*'s laps had also been at such speeds.

Reaching the end of the eastern leg for the second time, it took *Novelty* just one minute and seven seconds to slow down, turn round and race off towards the west, on the fourth leg of the thirty-five-mile journey. This was half the time of the previous turn-round – the champion was settling into her stride. Once again she dashed down the track towards the grandstand, moving as smoothly as silk. Some gentlemen cheered, others stood and waved their hats. But moments later there was a large puff of steam from the bottom of her furnace and *Novelty* coasted to a halt. It was just past two o'clock.

According to the judges report, 'the joints of the Boiler gave way', which they attributed to their newness. But Robert Stephenson, who wrote a report on the trials with Joseph Locke, disagreed. The failure was due not to a 'green joint' but to the collapse of a pipe that conveyed hot gases from the furnace to the boiler. They were confident in their assessment, 'having seen the pipe when it was taken out . . .'.

Regardless of the cause, the situation was hopeless. It would take several days to strip the boiler down, make good the repairs and reassemble the entire unit, resealing all the joints. Ericsson walked across to John Rastrick, the presiding judge, and gracefully withdrew from the competition. It would be said afterwards that the hot-headed Swede should not have capitulated so readily, but he had little choice. *Novelty* had broken down three times now, and there was no guarantee that the hastily constructed engine would not fail again. Besides, the contest had already dragged on into a second week and the judges could hardly be expected to extend it still further.

Magnanimity was not on Hackworth's agenda that day, and he presented the judges with a letter formally requesting another trial. The judges were unsympathetic to his demands:

> having considered the enormous consumption of fuel on the Tuesday's Experiment, and the general construction

> of his Engine we found we could not recommend it to the Directors' consideration as a perfect Engine, & therefore as it was also over-weight we did not think it necessary to spend further time in experimenting upon it.

Sans Pareil, for all her strong showing during her trial, possessed several anachronistic features that made her unsuitable for running on a main-line railway. Her high fuel consumption could have been improved by reducing the force of the blast pipe, but her old-fashioned boiler would never give the same fuel efficiency as a multi-tubular one. Another disadvantage of the flue-tube boiler, which may not have occurred to the judges, was that it limited the size that locomotives could attain. The Rainhill locomotives were all comparatively small, but the time was rapidly approaching when larger engines would be required for long distance travel.

The size constraint has to do with the relation between areas and volumes, referred to earlier. As engine size increased, the surface area of a flue tube became progressively smaller relative to the volume of water to be boiled. Multi-tubular boilers were not so constrained because, as boiler size increased, so did the number of tubes and with them the surface area available for heat transfer. Return-flue locomotives *could* be built larger than *Sans Pareil*, but their limited steaming capacity would not allow them to attain high speeds for long journeys. *Samson*, for example, built in 1838 for a Nova Scotia colliery, had a boiler that was twice as long as *Sans Pareil*'s, but she was a plodding workhorse, not a steeplechaser. *Sans Pareil*'s other shortcomings, aside from being overweight, included her vertically oriented pistons, which caused her to roll quite alarmingly when travelling at speed. Their orientation also effectively counteracted the action of any springs on the driving wheels. One of the original stipulations of the company was that the engine must be supported on springs. However, as Dixon wrote of *Sans Pareil* after the trials, 'as for being on Springs I must confess I cannot find them out . . . neither can I perceive any effect they have'. So, like the earliest locomotives, *Sans Pareil* was effectively without springs, and this would have been detrimental to track and wheels alike.

When the judges refused his request for another trial Hackworth began remonstrating again over the accuracy of the company's scales, but they had already been checked and found to be accurate. Dixon was most unsympathetic, describing Hackworth's his locomotive as 'very ugly' and concluding that he 'must return to his preaching and own that he has been taught a lesson of humility'.

While *Novelty* had been making her final attempt to win the prize, Timothy Burstall had been preparing *Perseverance* for her Rainhill debut. He had no chance of beating *Rocket*, as he had conceded to the judges, but he wanted to see what she could do. As it happens *Perseverance*, looking like a giant iron wine bottle slung between four wheels, could do very little. She wheezed and rattled down the track no faster than a person could walk, but it amused the crowd. Dixon was as uncharitable towards Burstall and his engine as he had been towards Hackworth:

> Burstall . . . upset his [locomotive] in bringing [it] from Liverpool to Rainhill and spent a week in pretending to remedy the injuries wherein he altered and amended some parts every day till he was last of all to start & a sorrowful start it was; full 6 Miles an hour cranking away like an old Wickerwork pair of Panniers on a Cantering Cuddy Ass.

With the contest officially at an end, *Rocket* was declared the winner. Nobody could deny that the Stephensons' locomotive had been the only one to satisfy all the requirements of the contest. Nevertheless, there were several accusations that the competition had been unfairly biased in favour of the Stephensons, and the complaints and recriminations carried on for many months to follow. Most of the critical press reports were heavily biased in favour of *Novelty*, which was hailed as the most advanced locomotive of the time. *The Liverpool Mercury* expressed its regret that *Novelty* was 'not built in time to have the same opportunity of exercising that Mr Stephenson's engine had, or that there is not in London, or its vicinity, any railway where . . . it could have been tried'.[9]

It congratulated Stephenson for the prize he was about to receive, but qualified the merit of his achievement:

> This is due to him for the perfection to which he has brought the old-fashioned locomotive engine; but the *grand prize of public opinion* is the one which has been gained by Messrs Braithwaite and Ericsson, for their decided improvement in the arrangement, the safety, simplicity, and the smoothness and steadiness of a locomotive engine; and . . . it is beyond a doubt – and we believe we speak the opinion of nine-tenths of the engineers and scientific men now in Liverpool – that it is the principle and arrangement of this London engine which will be followed in the construction of all future locomotives.

The popular *Mechanics' Magazine*, which reprinted *The Liverpool Mercury* article, echoed the sentiment. It too congratulated Stephenson on his success, noting that if

Parliament had seen fit to reward Harrison for the accuracy of his clock, the directors of the railway were equally correct in awarding Stephenson the £500 premium. However, it contended that *Sans Pareil* was 'at least as good' as *Rocket*, and that 'As the timekeeper of Harrison has been excelled by others . . . so may "The Rocket" be eclipsed by other locomotive engines, by the very engines against which it tried its powers on the present occasion . . .'.

The magazine finished its editorialising with an appeal to the Liverpool & Manchester Railway to 'grant a liberal indemnity to the competitors . . .'. Before going on to describe some of the competing engines, it published an article on the history of the Liverpool & Manchester Railway. This was to correct some errors that had appeared in an earlier account on the same subject, published in the previous edition. The latest article was written by an unnamed 'gentleman intimately acquainted with all the circumstances of the case'. The article reiterated that the line was originally surveyed by William James, and then resurveyed by George Stephenson. The point was made that Stephenson's measurements contained many serious errors, and these were principally to blame for the defeat of the railway Bill. 'This fact is notorious, and will be remembered by all those who attended to the parliamentary proceedings in 1825.' When the company engaged the Rennies, they 'sent down Mr Vignoles' to make a new survey, which was shorter than Stephenson's. After a disagreement between the directors and the Rennies, 'Mr Stephenson was called upon to lay down the railway on the line surveyed by Mr Vignoles.' There is little doubt that the author of this pointed criticism of George Stephenson was Vignoles himself. He is also likely to have influenced, if not written, the articles predicting *Rocket*'s eclipse.

Vignoles, a long-time champion of Novelty, *penned this rather imaginative drawing of how the engine might be used to convey passengers.*

Regardless of these minority reports, *Rocket*'s dazzling success at Rainhill removed any lingering doubt among the directors about the future of locomotives. Before the month was out they had purchased *Rocket* and placed an order for four more similar engines from Robert Stephenson & Co., to be delivered within three months.

Although George Stephenson had silenced his critics, he still had a few enemies on the board, notably James Cropper. Cropper had always been vehemently opposed to Stephenson and to locomotives, and had led the charge for stationary engines from the outset. Now that locomotives had become universally acceptable, he redirected his animosity towards the Stephensons by throwing his support behind Messrs Braithwaite and Ericsson and their alternative concept of locomotive design. Once *Novelty* had been repaired it was tried out again at Rainhill, but it was not a success and continued to break down. In spite of this, and under Cropper's influence, the directors placed an order with Braithwaite and Ericsson for two similar but larger locomotives, much against George Stephenson's advice. Named *William IV* and *Queen Adelaide*, in honour of the new king and his wife, they underwent trials on 18 February 1831. This was just one week before the trial of a new locomotive, named *Samson*, from the Forth Street Works.

Both trials were reported in the press, and the *Manchester Guardian*, while glowing in its praise of *Samson*, was very critical of *William IV*:

> the William started from Manchester station . . . with a load of 5 waggons, the gross weight of which was 20 tons 6 cwt 1 qr. After getting fairly in motion, the William proceeded for about two miles at a speed of about 10 miles an hour, which however gradually declined to about 6 or 7, and occasionally to less than 5 miles an hour, and it was very obvious that the engineer *could not succeed in keeping up an adequate supply of steam*. At the conclusion of the first hour the William had gone exactly 8 miles and a half, the greater part of which was on a descent . . . [the newspaper's emphasis]

Progress was so painfully slow that on two occasions one of Stephenson's locomotives had to assist *William IV* with a push, shunting her on to a siding so the regular trains on the line could pass.

Samson, which was one of the first Stephenson locomotives fitted with a cranked driving axle, was a much larger and more powerful engine, and hauled a load of over 150 tons at about 20 mph. The article concluded that:

> We imagine that the performance yesterday, will, at any rate, put an end to the system of petty detraction, which has been so long and so incessantly levelled at Mr Stephenson and his engines, by a little knot of pseudo-mechanics,* who appear to have imagined, that they had

> * Of course we do not include in this term Captain Ericsson, who is unquestionably a man of considerable mechanical talent.

> an interest in deprecating every thing he did, and in crying up even the blunders and the failures of other parties. We are very curious to see how this, and some other recent occurrences, will be treated in the Mechanics' Magazine, the organ of the party to which we have alluded.

The *Mechanics' Magazine*, which reprinted large sections of the *Manchester Guardian* article, denied the accusation that it had ever been 'an instrument of 'petty detraction' towards either Mr Stephenson, or any other individual . . .'. However, it did admit that:

> We may not have expressed so high an opinion of the works of Mr [George] Stephenson's hands as others have done – and no wonder, since nothing less will content his partizans [*sic*] than to rank him with the Watts and Davys of the age!

Braithwaite and Ericsson, much to their credit, never entered the fray, and were most courteous in all their dealings. Although things did not work out well for them with their locomotives, they wrote to the directors of the Liverpool & Manchester Railway expressing their gratitude 'for the attention we have experienced during the experiments, which have been facilitated by every assistance on the part of the Company'. Referring to the engines many years later, Ericsson wrote, 'They proved utter failures for want of steam . . .'.

Hackworth, in contrast, was far from conciliatory, and felt he had been ill used at Rainhill. His first grievance was over the poor workmanship of the riveting of his boiler. He was also smarting from not being allowed a second trial of his engine, and it must have rankled that *Novelty* had been allowed three attempts. His situation was made all the more difficult by the conviction that his locomotive was second to none, and that its failure was attributable to bad luck and to the failings of others rather than to its pedestrian design. His bad luck had not ended with the faulty feed pump, either. Some time after his trial – it is not clear exactly when – he discovered a crack in one of the cylinder castings, manufactured, as already mentioned, by Robert Stephenson & Company. Robert Young, Hackworth's grandson and author of his biography, claimed the cracked casting was responsible both for *Sans Pareil*'s breakdown at Rainhill and for her inordinately high fuel consumption. But this is entirely unfounded, and there is no question that the breakdown, initiated by the feed pump, was entirely due to the melting of the fusible plug.

Regardless of how blind Hackworth was to the inherent shortcomings of his own locomotive, the perception that fair play had not prevailed at Rainhill was not restricted to the losers and their supporters. William Kitching, one of the board members of the Stockton & Darlington Railway, wrote Hackworth a letter of commiseration:

> I should have been extremely glad to have heard that fair play had been allowed to the different engine makers who were competitors for the premium; from the very cool reception that a Timothy received from Forth Street, George showed very plainly that he was much afraid of the Shildon production; and well they might, as they well knew that every engine which they have yet sent has had great need of mending . . . From what I have heard of the doings with you, it appears that the engine which had a Booth as the inventor of the copper pipes in the boiler, was, without either judge or jury, to be the winner'.

Once the cracked cylinder casting had been replaced, Hackworth conducted a demonstration of his locomotive for the Liverpool & Manchester Railway, though it appears from the letter he wrote to the directors that none of them was sufficiently interested to witness the proceedings:

> You are doubtless aware that on a recent occasion the Locomotive Engine *Sans Pareil* failed the task assigned to her by the Judges . . . suffice it to say that neither in material construction nor in principle was the Engine deficient, but circumstances over which I could not have any control . . . compelled me to put that confidence in others which I found with sorrow was but too implicitly placed . . . The whole alteration which has been made is the removal of a cylinder which failed from its defective casting –
>
> Having reached Rainhill on Wednesday . . . the Engine was moved a short distance on your Railway – I felt peculiarly anxious to have tested my Engine over the same ground as that on which the prize trial took place – finding that impossible . . . I was confined to 3/4 of a mile . . . notwithstanding the Rails were very dirty and the distance was but short a speed of 30 miles per hour was attained – On the following morning . . . one of your own Agents weighed in 7 cwt. of Fuel – the Engine was kept continually in motion . . . till 6 at night . . . during this day the Engine ascended and descended the Inclined Plane repeatedly, with water, fuel and 18 Passengers . . .
>
> I felt peculiarly anxious that you gentlemen, or some of you should have been Witness of the performance of my Engine – believing that you would have been well satisfied that I have not formed an exaggerated opinion of

> its merits when I state that in Simplicity, Power and economy of fuel it is a Machine decidedly unequalled by anything of the kind yet seen – I should not have the least hesitation in having *Sans Pareil* placed on your line of Railway, by the side of every Engine yet tried . . . and engaging to forfeit the Engine – if it did not in every way exceed anything performed on the day of the trial . . . I feel myself injured in some reports printed and verbal which have circulated – of any participation in this I entirely acquit your board – Your offer to purchase the Engine I accept with thanks – but you will pardon me when I honestly add that £550 [£32,000] does not by any means compensate me for the Expenses and labour bestowed.

At their first meeting after the trials the directors agreed to purchase *Sans Pareil* for £550, once the cracked cylinder had been replaced. The locomotive was not intended for the Liverpool & Manchester Railway but was for use on a subsidiary branch line, the Bolton & Leigh Railway, where it remained in regular service for the next fifteen years. After that, it was used as a stationary engine for another sixteen years. During all those years it remained largely unaltered, the main change being that the original wooden spoked wheels were replaced by cast iron ones. The locomotive can still be seen today at the National Railway Museum, York. It looks remarkably intact and in good condition, in marked contrast to *Rocket*.

At the same directors' meeting a proposition was read from Mr Burstall offering to build a locomotive for the company: Burstall's audacity, it appears, knew no bounds. The directors postponed making any decision until the judges had delivered their report of the trials. The rest of the meeting was taken up with matters pertaining to the completion of the line. On the recommendation of George Stephenson, *Rocket* was to be taken to Chat Moss and used for hauling ballast for the construction of the line. The judges' report was delivered and read at a meeting the following day, after which it was unanimously agreed to award the premium to Messrs Booth, Stephenson and Stephenson. While Burstall's proposal was not accepted, the directors did award him £25 (£1,500) to help defray his Rainhill expenses.

CHAPTER 11

Triumph and tragedy

The year 1830 began bitterly cold. The Reverend Leonard Jenyns, the naturalist who declined the voyage aboard the *Beagle* that Darwin subsequently accepted, described how pitchers of water froze and cracked in the bedrooms of his Cambridgeshire home, and how the saliva of draught horses froze into six-inch icicles. The frost and snow continued into February and it was so cold that much of the Thames became completely frozen. There was some respite in March, but the sleet and snow returned for much of April, and a stormy and unsettled May was followed by four months of rain.

The appalling weather was as dismally depressing as the political situation. The Duke of Wellington had been the nation's hero when he conquered Napoleon at the battle of Waterloo, but as Prime Minister he was widely despised. After so many years of harsh repression his Tory party had began to liberalise its treatment of the downtrodden working classes. But the Iron Duke nipped all that in the bud when he became leader in 1828. He was staunchly opposed to parliamentary reform, and saw no reason to change a system which kept the right to vote from the masses, and which protected the vested interests of the privileged few with a parliament of their peers. Of all the working classes, the agricultural labourers were among the poorest paid, and some of their more militant members demonstrated their dissatisfaction by setting fire to hayricks. There had been hayrick burnings in the past, but towards year's end the number of conflagrations had reached almost epidemic proportions.

The liberalisation movement had got under way in 1822, following the death of Lord Castlereagh, the Tory hardliner who had established a militia for maintaining domestic law and order. His successor as Foreign Secretary and Leader of the House of Commons was George Canning, a moderate who became Prime Minister five years later. Among the other Tory moderates was William Huskisson, the Member for Liverpool. Huskisson was one of Britain's leading economists, and played a major role in drafting trade legislation, some of which laid the foundations for Britain's free trade policy. Paradoxically, he favoured protecting British producers against cheap grain imports and was accordingly given responsibility for legislating the Corn Law, passed by Parliament in 1815. Under this law, duty-free foreign grain was permitted to be imported only when domestic prices were above a certain level. While protecting the profits of the landowners, the Corn Law, by maintaining elevated grain prices, kept bread prices high, hurting the working classes. Notwithstanding his advocacy of the Corn Law, Huskisson was a strong voice for free enterprise, and a keen supporter of the Liverpool & Manchester Railway.

In later years Huskisson began having doubts about the

Corn Law, and proposed new measures to reform the legislation. His proposal was opposed by the Duke of Wellington, and by the other hard-liners among the Tories, and when the Duke became Prime Minister, Huskisson was relegated to the back benches. Shunned by Wellington and his conservatives, Huskisson became one of the leading proponents of parliamentary reform.

George Stephenson had more pressing things on his mind in 1830 than politics – with the 15 September opening looming ever closer, he had a railway to finish. The four locomotives the company ordered from Forth Street, named *Meteor*, *Comet*, *Dart* and *Arrow*, incorporated several improvements over *Rocket.* The cylinders were placed more nearly horizontally, minimising the lurching movements. By reducing the diameter of the boiler tubes from 3 in. to 2 in., and increasing their number from twenty-five to eighty-eight, the heating area was more than doubled for the same size boiler. The greatly improved steaming capacity permitted the cylinder bore to be increased from 8 in. to 10 in. This enlarged the surface area of the piston by more than 50 per cent, increasing the force generated by the steam in the same proportion. The new engines were therefore capable of hauling greater loads at higher speeds.

Two more locomotives were ordered in February 1830. The first of these, *Phoenix*, was delivered in June, while the second, *North Star*, arrived two months later. The length of the boiler had been increased by 6 in., but no information is available on the size and number of boiler tubes. The cylinders were increased in diameter to 11 in., increasing the surface area of the pistons by about 20 per cent. The most significant improvement over the previous locomotives was the provision of a separate smokebox at the front of the boiler, for collecting the smoke and exhaust gases from the boiler tubes and delivering them to the chimney. In *Rocket*, and its immediate descendants, this function was served by the base of the chimney being broadened to cover the front of the boiler tubes. The front of the smokebox had two doors: an upper one to give cleaning access to the tubes, and a lower one for the removal of the ash that collected at the bottom. The performance of these two engines was little different from that of the previous four, but the smokebox made the boiler tubes easier to clean and maintain.

An eighth locomotive, *Northumbrian*, was delivered in August, and this was the most advanced engine in the world. The boiler tubes had been decreased in diameter to 1⅝ in., and increased in number to 132. This more than trebled the heating area over that of *Rocket*, for only a 6 in. increase in boiler length. The firebox was enlarged too, increasing the heating surface to about 36 sq. ft, some 40 per cent greater than *Rocket*'s. Furthermore, the firebox was built as an integral part of the boiler rather than a separate add-on. *Northumbrian* was the fastest and most powerful locomotive, and could haul a fifty-ton load up the incline approaching Rainhill at 7½ mph.

Although George Stephenson was busy, he still made time to entertain certain guests with special trips along finished sections of the line. This was a deliberate plan on the part of himself and the directors, to popularise rail travel among people of influence. Thomas Creevey was one of the first such guests. He was the MP for the rotten borough of Downton, Wiltshire. Along with his ally Lord Sefton, he had been one of the most vocal opponents of the railway. After his railway ride, Creevey recounted that:

> I had the satisfaction, for I can't call it pleasure, of taking a trip of five miles in it, which we did in just a quarter of

> an hour – that is 20 miles an hour . . . during these five miles, the machine was occasionally made to put itself out or *go it* and then we went at the rate of 23 miles an hour, and just with the same ease as to motion or absence of friction as the other reduced pace. But the quickest motion is to me *frightful*; it is really flying, and it is impossible to divest yourself of the notion of instant death to all upon the least accident happening. It gave me a headache which has not left me yet. Sefton is convinced that some damnable thing must come of it . . . Altogether I am extremely glad indeed to have seen this miracle, and to have travelled in it . . .

Stephenson appears to have won over one of his most ardent detractors.

On the morning of 14 June 1830, the directors assembled at Liverpool's Crown Street station for a special return train trip to Manchester 'to inspect the Works, and prove the capabilities of the Locomotive Engine'. *Arrow*, driven by George Stephenson, was hitched to two passenger carriages, to which seven wagons were added, loaded with stone blocks, for a total load of thirty-three tons. At the bottom of Whiston incline *Arrow* was met by *Dart*, to assist with the gradient. When the two engines reached the level section at Rainhill, *Dart* was uncoupled and *Arrow* carried on alone, travelling at 16 mph. They reached Manchester just after eleven o'clock, having stopped twice along the way to take on water. After inspecting the warehouses and other buildings under construction, the group repaired to an office where they passed a resolution: 'expressing their strong sense of the great skill and unwearied energy displayed by their Engineer, Mr George Stephenson, which have so far brought this great national work to a successful termination . . .'. The seven wagons were unhitched for the return journey, and the directors were whisked back to Liverpool at speeds of 25 mph and more.

Of all the special visitors to the railway prior to its opening, the one who delighted George the most was the enchanting Fanny Kemble. Miss Frances Anne Kemble (1809–93) was appearing on stage in Liverpool at the time, and the directors had invited her, along with fifteen other guests, to take a trip aboard a train drawn by *Northumbrian*. Three carriages were hitched to the locomotive, and after spending the first part of the journey with the other guests, the actress was invited to ride on the footplate beside George Stephenson. The vivacious twenty-year-old had never ridden a locomotive before, but gave a remarkably detailed account of the experience:

> We were introduced to the little engine which was to drag us along the rails. She (for they make these curious little fire horses all mares) consisted of a boiler, a stove, a platform, a bench, and behind the bench a barrel containing enough water to prevent her being thirsty for fifteen miles, – the whole machine not bigger than a common fire engine. She goes upon two [pairs of] wheels, which are her feet, and are moved by bright steel legs called pistons; these are propelled by steam, and in proportion as more steam is applied to the upper extremities (the hip-joints, I suppose) of these pistons, the faster they move the wheels; and when it is desirable to diminish the speed, the steam, which unless suffered to escape would burst the boiler, evaporates through a safety valve into the air. The reins, bit, and bridle of this wonderful beast, is a small steel handle, which applies or withdraws the steam from its legs or pistons, so that a child might

Miss Frances Anne Kemble.

> manage it. The coals which are its oats, were under the bench, and there was a small glass tube affixed to the boiler, with water in it, which indicates the fullness or emptiness when the creature wants water, which is immediately conveyed to it from its reservoirs. There is a chimney to the stove but as they burn coke there is none of the dreadful black smoke which accompanies the progress of a steam vessel. This snorting little animal, which I felt rather inclined to pat, was then harnessed to our carriage, and . . . we started at about ten miles an hour.
>
> The steam horse being ill adapted for going up and down hill, the road was kept at a certain level, and appeared sometimes to sink below the surface of the earth and sometimes to rise above it.

The first cutting they passed through was Olive Mount, a narrow ravine, over a mile long and up to seventy feet deep, hacked and blasted out of solid rock.

> You can't imagine how strange it seemed to be journeying on thus, without any visible cause of progress other than the magical machine, with its flying white breath and rhythmical, unvarying pace, between these rocky walls, which are already clothed in moss and ferns and grasses; and when I reflect that these great masses of stone had been cut asunder to allow our passage thus far below the surface of the earth, I felt as if no fairy tale was ever half so wonderful as what I saw. Bridges were thrown from side to side across the top of these cliffs, and the people looking down upon us from them seemed like pigmies standing in the sky.

Olive Mount, the first cutting after leaving Liverpool.

The material excavated from Olive Mount was used to build a huge embankment across the succeeding valley. Named the Roby embankment, it was about two miles long, varying in height between 15 ft and 45 ft.

As *Northumbrian* burst clear of Olive Mount her astounded passengers, having spent the previous several minutes looking up at the trees and sky, suddenly had their world turned upside down as they gazed down upon treetops and a vast expanse of countryside. After Roby embankment the line crossed the turnpike road, and Stephenson would have slowed down to make sure no stagecoaches or other traffic was approaching. Passing the village of Huyton on their left, they soon began ascending Whiston incline towards Rainhill, but *Northumbrian* would have made light work of the gradient with such a light load to haul. They sped along the level section at Rainhill, then raced down the Sutton incline.

Two miles further on and they approached the most spectacular architectural feature of the entire route, the majestic Sankey viaduct, spanning the valley of the same name 70 ft below. As the passengers gazed down in awe at the Sankey canal, with its Lilliputian barges, Stephenson's hand reached for the regulator valve. Cutting off the steam to the cylinders, the train coasted to a gentle halt. At this point the greying engineer jumped down on to the track, helped the actress down, and the two set off together along the line. It was a stiff walk down the steep embankment to the fields below, but what a magnificent view they had of the massive brick arches, all nine of them, that carried the railway across the valley. And it was all his own work: the viaduct, the line, winning acceptance for the locomotive. George Stephenson must have felt as tall as the towering columns as the beautiful young woman at his side admired his work.

> He explained to me the whole construction of the steam engine, and said he could soon make a famous engineer of me, which, considering the wonderful things he has achieved, I dare not say is impossible. His way of explaining himself is peculiar, but very striking, and I understood, without difficulty, all that he said to me. We then returned to the rest of the party . . .

Soon after leaving Sankey they passed the old town of Newton, and clattered across the four-arched Newton viaduct, which spanned a turnpike road and a brook. *Northumbrian* was now racing along, and may have reached speeds never before experienced by humans. According to Fanny Kemble the locomotive was travelling:

> at its utmost speed, 35 miles an hour, swifter than a bird flies . . . You cannot conceive what that sensation of cutting the air was; the motion is as smooth as possible, too. I could either have read or written; and as it was I stood up, and with my bonnet off 'drank the air before me.' The wind, which was strong, or perhaps the force of our own thrusting against it, absolutely weighed my eyelids down. When I closed my eyes this sensation of flying was quite delightful, and strange beyond description; yet strange as it was, I had a perfect sense of security, and not the slightest fear.

What a contrast to the terrified Thomas Creevey's recollections.

She went on to describe how the power of the locomotive was demonstrated by hitching a second train, and a laden timber wagon:

> this brave little she-dragon of ours flew on . . . when I add that this pretty little creature can run with equal facility either backwards or forwards, I believe I have given you an account of all her capacities.

Stephenson may have allowed himself a complacent smile as they approached Chat Moss, a flat and desolate place which seemed to go on for ever. During the first survey of the line across the trackless peat bog, William James, a rather portly gentleman, found himself sinking into the mire. With great presence of mind he threw himself on to his side and managed to roll over and over until he found firmer ground. John Dixon, who helped construct the railway, was less fortunate. Visiting the bog for the first time, he slipped off one of the planks that had been laid down for walking upon, and quickly sank up to his knees. Struggling to free himself only made matters worse, and he was soon up to his waist in the spongy wetness. Had it not been for the swift intervention of the workmen he might have disappeared for ever.

Most people said it was impossible to lay a line across the quagmire where livestock feared to tread, but Stephenson thought otherwise. Following the advice of locals, he laid down a raft of wicker hurdles, branches, heather, hedge cuttings and the like, to form a floating embankment for the railway. But as fast as material was added, so it disappeared. Recounting his experience many years later, Stephenson recalled that after several weeks of labour there was not 'the least sign of our being able to raise the solid embankment one single inch . . .' saying that they 'went on filling without the slightest apparent effect'. Stephenson's assistants, as well as the directors, thought the situation hopeless, and a meeting was held to decide whether he

Laying wicker hurdles to form the embankment across Chat Moss.

should carry on with the seemingly impossible task. But Stephenson, who 'never for one moment doubted' he would succeed, was allowed to continue.

Fanny Kemble did not confine her recollections of the adventure to the journey itself:

> Now for a word or two about the master of all these marvels, with whom I am most horribly in love. He is a man from fifty to fifty-five years of age; his face is fine, though careworn, and very original, striking, and forcible; and although his accent indicates strongly his north country birth, his language has not the slightest touch of vulgarity or coarseness. He has certainly turned my head.

> Four years have sufficed to bring this great undertaking to an end. The railroad will be opened upon the fifteenth of next month. The Duke of Wellington is coming down to be present on the occasion . . . The whole cost of the work . . . will have been eight hundred and thirty thousand pounds [about £51 million]; and is already worth double that sum. The Directors have kindly offered us three places for the opening, which is a great favour, for people are bidding almost anything for a place, I understand.

The day before the grand opening, William Huskisson, the Honourable Member of Parliament for Liverpool, delivered a stirring speech to a capacity crowd at the Liverpool Exchange. His enthusiastic audience was delighted to see that his health had been much restored. Huskisson, who was now in his sixty-first year, had been so severely indisposed during the previous election that he had been unable to make a single appearance in his home riding. His loyal supporters re-elected him anyway, for which he now gave his sincere thanks. He went on to tell them of the pending visit 'of a distinguished individual of the highest station and influence in the affairs of this great country . . .'. Without mentioning the Duke of Wellington by name, he assured his audience that the visit would convince him of the great importance of Liverpool to the economic good of the nation. The conclusion of his speech was met by the heartiest cheering that had 'ever burst from the lips of a Liverpool assembly'.

Visitors had been streaming into Liverpool to witness the opening of the railway since the end of the previous week. Fanny Kemble, chaperoned by her mother and accompanied by several friends, did not arrive until the night before.

By that time all the inns and hotels in the town were overflowing, and carriages stood in the streets because there was no more room in any of the stable yards. Fortunately the hotelier of the Adelphi was an old friend of the family, and was able to accommodate them in a tiny garret.

According to *The Liverpool Mercury*, the town had never before witnessed 'such an assemblage of rank, wealth, beauty and fashion'. But the event was not the prerogative of the rich, having captured the imagination of 'all classes of the community'. *The Liverpool Courier* maintained that 'the lower orders' were so 'absorbed in the anticipation of the day' that if a holiday had not been taken for granted, few would have shown up for work anyway.

The organisers must have breathed a sigh of relief when they awoke on that 15 September morning to find such a promising day after all the rain of the previous few months. People were up and about at first light, converging on any and every vantage point along the route that would give them a view of the procession. They crowded bridges, swarmed embankments and stood shoulder to shoulder and several deep beside the track. They even sat on rooftops, seemingly oblivious to any concern for safety.

Similar crowds had converged on Manchester, though they did not need to take their places quite so early, as the procession was not due until early afternoon. Thousands more gathered in fields along the way, many of them having arrived on foot the previous day and camped overnight. People arrived on horseback, and by a remarkable array of vehicles, from farm carts and tinkers' caravans to stagecoaches and stately carriages. Grandstands had been erected at several locations *en route*, including both terminals and the Sankey viaduct, and there were booths at every spot with a commanding view along the way.

Few events had ever attracted so many people, and while the chance to witness a cavalcade of the latest locomotives was reason enough for most to attend, the newspapers maintained the Duke's appearance had much to do with the record turnout. While some visitors doubtless wanted to see the famous man, others had come to demonstrate their disaffection with the Prime Minister, and his policies. Anticipating trouble from the radicals, the authorities had arranged for a significant presence of law enforcers, both civilian and military.

The Duke was to ride in a special carriage – over 30 ft long and 8 ft wide – drawn by *Northumbrian*. It was an ostentatious creation: all gilt mouldings and laurel wreaths, canopied by a rich crimson fabric supported on eight gilded columns and surmounted by the ducal coronet. Two somewhat lesser carriages, one in front and one behind, were to convey the other dignitaries of rank, preceded by an open carriage with the musicians of the Wellington Harmonic Band. The other guests, numbering 672 in all, were to be accommodated in carriages drawn by seven other locomotives.

The engines were marshalled in the company's Crown Street yard, where they were hitched up to their respective carriages. So guests could find their reserved seats more easily, each locomotive flew a flag corresponding to the colour of the passengers' tickets: *Phoenix* (dark green), *North Star* (yellow), *Rocket* (light blue), *Dart* (purple), *Comet* (deep red), *Arrow* (pink) and *Meteor* (dark brown). George Stephenson would drive *Northumbrian*, and, to give the other guests the opportunity of viewing the Duke and his entourage, his train would use the right-hand track, while the other seven locomotives used the other.

The guests began arriving before nine o'clock and the busy yard soon became the centre of much excitement and speculation as more carriages drew up and set down their distinguished passengers. Security was provided by a contingent of the 4th Regiment, while the band aboard the Duke's train provided a musical accompaniment. Meanwhile, the band of the King's Own Regiment played musical airs in an adjoining area, while a third band played on a stage, constructed on the premises of the adjoining William the Fourth hotel.

Shortly before ten o'clock the discharge of a heavy gun reverberated through the town. This was immediately followed by loud cheering, signifying the arrival of the Duke. His grace was an austere man in his early sixties, dressed completely in black, in mourning for King George IV, who had died in June. A loose Spanish cloak covered most of his person, save for a large medal he wore on his left breast. He was accompanied by his friends, the Marquis and Marchioness of Salisbury, with whom he was staying. The Duke looked remarkably well.

When the Marquis's carriage swung into the yard all three bands struck up with 'See, the conquering hero comes'. The crowds added their approbation with their spirited hurrahs and applause. The warm welcome gave his grace no small amount of satisfaction, which he acknowledged to the crowd. And as the bands played and the crowds cheered and waved, the Duke made his way to his carriage, taking his seat at the front. Among the many other honourable guests in his carriage were Prince Esterhazy, the Austrian ambassador, Count Potocki, the ambassador of Russia, Sir Robert Peel, the Home Secretary, Lord Wharncliffe, one of the Grand Allies, and William Huskisson. There were eighty guests aboard the Duke's train, including the directors, making a grand total of 752 in the eight trains.

At 10.40 a.m. precisely a signal gun was fired and the locomotives, led by *Northumbrian*, started off. The eight trains moved slowly at first, barely exceeding a walking pace. Excited spectators lined the route as far as the eye could see, and in their enthusiasm they had spilled on to the track. The civilian force policing the line could do nothing to persuade them to clear the way. Then the locomotives gathered speed and 'went forward with arrow-like swiftness, and thousands fell back . . .'.

Opening-day of the Liverpool & Manchester Railway; the Duke of Wellington's train is on the left.

Some of the men in charge of the locomotives were as distinguished as many of the guests aboard their trains. Leading the seven engines on the left-hand track was Robert Stephenson aboard *Phoenix*, with his uncle of the same name behind him driving *North Star*. *Rocket* was driven by Joseph Locke, protégé of George Stephenson and good friend of Robert – he would eventually become President of the Institution of Civil Engineers. The locomotives' crews were smartly turned out, as reported in *The Liverpool Courier*: 'the men and boys who worked them [were] dressed in the new blue livery of the company . . . the words "fireman", with the name of the machine to which they were attached, appearing on their caps, in red letters . . .'. Whether the distinguished drivers were dressed in company livery is unclear. Perhaps only their firemen were so attired.

By the time the trains entered Olive Mount cutting they were travelling at about 24 mph. People crowded the overhead bridges and even stood on the jagged rocks that projected from the walls of the great chasm. And then the cavalcade was thundering across open countryside.

> Vehicles of every description stood in the fields on both sides, and thousands of spectators still lined the margin of the road . . . Under Rainhill Bridge, which, like all the others, was crowded with spectators, the duke's car stopped until several of the other trains passed, thus affording the passengers a most excellent opportunity of seeing the whole of the noble party.

Soon *Northumbrian* was under way again, catching up with the other trains and taking the lead once more. It then pulled so far in front that it could be seen by some of the passengers in the other trains:

> gliding beautifully over the Sankey Viaduct, from which a scene truly magnificent lay before us. The fields below

> were occupied by thousands, who cheered us as we passed over the stupendous edifice; carriages filled the narrow lanes and vessels in the water had been detained in order that their crews might gaze up at the gorgeous pageant passing far above their mast-heads. Here again was a grand stand, and here again enthusiastic plaudits almost deafened us.

Shortly after crossing Sankey viaduct the cavalcade arrived at Parkside, seventeen miles from their Liverpool starting point, and the locomotives stopped to take on water. It was somewhat before noon, the journey having so far taken just under an hour. *Northumbrian*, with the Duke's party aboard, had come to a halt, and *Phoenix*, leading the seven trains on the other track, had pulled ahead some 300 yards to refill its water barrel.

Taking the opportunity, some of the gentlemen aboard the Duke's train clambered down from their carriages to stretch their legs and were milling about in the narrow gap between the two tracks, socialising in the festive spirit of the occasion. Spirits were exuberantly high, 'all hearts bounding with joyous excitement'. Mr Huskisson was in conversation with Mr Joseph Sandars, the founding director of the railway, and was congratulating him on bringing the magnificent work to such a triumphant conclusion. 'Well, I must go and shake hands with the Duke,' said Huskisson taking his leave, adding that this seemed an appropriate day for doing so. Huskisson's decision to use the auspicious occasion to patch up his differences with the Duke may have been entirely spontaneous, but it was reported that the reconciliation was initiated by William Holmes, a fellow MP, who thought 'that these two statesmen ought never to have been separated'. In any event, Huskisson made his

This illustration of a locomotive taking on water at Parkside was painted the year after the opening of the railway.

way back to the Duke's carriage, and, on seeing him standing outside, his grace extended his hand over the side. The two men shook hands most cordially and exchanged compliments. Their conversation might have continued, but there was a sudden panic aboard the Duke's carriage: several of the passengers had noticed a train approaching on the other track. It was drawn by *Rocket*. 'Get in, get in!' they cried to the men standing outside the carriage, galvanising them into frenzied action. Mr Littleton, the MP for Staffordshire, scurried up the narrow step ladder to the Duke's carriage, then managed to pull Prince Esterhazy up behind him. Holmes, who was standing beside Huskisson, tried to do the same, but failed. Realising the futility of a second attempt, he flattened himself against the side of the carriage.

There was a 4 ft gap between the tracks but the Duke's extra-wide carriage extended 2 ft into this space and the approaching locomotive projected beyond its own track too, leaving about 18 in. between the two vehicles. If the two gentlemen remained calm and pressed themselves tightly against the Duke's carriage they would be perfectly safe. But Huskisson seemed unable to control himself.

'For God's sake! Mr Huskisson,' cried Holmes, attempting to steady his companion, 'be firm!' But Huskisson was not listening. He hesitated for a moment, staggered as if completely confused, then made a frantic dash for the open carriage door.

Huskisson was not a fit man. Neither was he very steady on his feet, because of a partial lameness in one of his legs. He therefore had little chance of reaching the door and hauling himself into the carriage in time, regardless of the many willing hands there to help him.

Joseph Locke, at *Rocket*'s controls, saw the men scurrying for safety directly they saw him approaching. But it was impossible to bring the locomotive to a halt before it reached *Northumbrian*. All he could do was hope. And pray.

The door of the ducal carriage was so wide that it extended well into *Rocket*'s path. A collision was inevitable and when the locomotive struck it shattered the edge of the door, ripping off a strip of the decorative red baize. Whether Huskisson was struck by the flying door or whether he merely tripped and fell is unclear, but he landed, face down, in the narrow space between the two trains. As he fell, his left leg was flung on to the line, and an instant later it was pulverised beneath a passing wheel. The wheel passed obliquely across his thigh, 'squeezing it almost to a jelly' and breaking his left femur in two places. In doing so it 'laid the muscles bare from the ankle, nearly to the hip, and tore out a large piece of flesh as it left him.' Blood gushed from the mutilated leg and Huskisson uttered a faint scream. His wife, who had witnessed the accident along with several other ladies aboard the Duke's carriage, gave a shriek of horror. Nobody who heard the terrible scream would ever forget it.

With some difficulty the Earl of Wilton and Mr Joseph Parkes of Birmingham managed to lift Huskisson from the ground.

'This is the death of me!' exclaimed the injured man.

'I hope not, sir,' replied Parker.

'Yes, it is.'

Several of his friends were now by his side and Huskisson took one of them, named Staniforth, by the hand. 'I am dying,' he told him. 'Call Mrs Huskisson.' This was said 'in the most cool and collected manner [and he] gave such directions as he thought best fitted the situation . . .'.

Mr Staniforth took off his frock coat and rolled it into a ball as a pillow. But when he saw Mrs Huskisson approach he threw the coat over his friend's injured leg to conceal the dreadful wound. The distraught woman threw herself upon her husband and began kissing him. But she was soon obliged to leave his side so that two medical gentlemen, Drs Brandreth and Southey, both from London, could attend him. Placing a tourniquet around the upper part of his thigh, they managed to stem the bleeding, and had him carried to the side of the track on a makeshift stretcher. They were soon joined by Mr Hensman, a surgeon. After a joint consultation they decided he should be conveyed with all possible speed to Manchester, where his injuries could best be attended.

As *Northumbrian*, the fastest of the locomotives, was at the front, she would make the journey, using the musicians' open carriage as a temporary ambulance. The rest of the train was uncoupled, the musicians scrambled out, and Huskisson, now unconscious and looking 'pale and ghastly as death', was gently laid down on the floor of the carriage. With his wife 'hanging over him in an agony of tears', and accompanied by the doctors and a small group of close friends, the train sped off towards Manchester. George Stephenson was driving, and the train reached speeds approaching 40 mph, probably the fastest any humans had ever travelled.

The Duke wanted to cancel the rest of the opening ceremonies and have the entire procession return to Liverpool immediately. The directors acceded to his grace's wishes and word was spread that they were going to turn back. But the news was not well received by most of the other participants. Some of the investors suggested that failing to complete the ceremony could have a negative impact on the entire project. This point was not missed on the directors, and they made an appeal to the Duke, but he remained steadfast, and was joined in his intransigence by Sir Robert Peel. The reeves of Manchester and Salford respectfully pointed out that huge crowds were awaiting their arrival, and if they were disappointed there was no telling what consequences might befall the town. 'There is something in that,' remarked the Duke, who had listened, stony-faced, to their appeal. The reeves had apparently struck a sympathetic chord with the Duke's commitment to law and order.

After a lengthy discussion between Wellington and the directors it was agreed that the procession could continue. There was a problem – the Duke's train was without an engine. There was no way of shunting the carriages on to the other line and no other locomotive was available. So they had to improvise. The trains hauled by *Phoenix* and *North Star* were coupled together, and a long chain was passed across the track to the front of the Duke's carriages. The makeshift arrangement worked remarkably well, and, after a delay of an hour and a half, the procession set off again.

The crowds lining the route, completely unaware of Huskisson's accident, cheered and waved as the cavalcade went past. If they were puzzled that their greetings were not returned by the grim-faced passengers, it did not diminish their enthusiasm. Chat Moss, normally desolate and foreboding, was dotted with groups of merrymakers, many of them gathered around camp fires on which they were cooking their provisions. Flasks of good cheer were passed from one to another to drink to the success of the new railway. Their festive mood was a stark contrast to the gloom aboard the passing trains.

Before it reached Manchester the procession was rejoined by *Northumbrian*. Stephenson had good news: Huskisson seemed to be doing well. There was a brief pause while *Northumbrian* was recoupled to the Duke's short train, then they were off again.

When *Northumbrian*, with the injured Huskisson aboard, had reached Eccles, some three miles outside Manchester, the physicians decided he would be better off in a private house. He was therefore taken to the home of the Reverend Blackburne, the vicar of Eccles, where he was made as comfortable as possible. There was some discussion among the physicians about amputating the mangled leg, but they decided against it because of his weak condition. Even if he had been strong enough to face the terrible ordeal, the outcome would probably have been the same. Medicine was in a primitive state of ignorance at the time. There were no anaesthetics, and surgical procedures were an agonising ordeal, ameliorated only by alcohol – with the negative side-effect of increased bleeding – or by administering laudanum (opium dissolved in alcohol). There were no antiseptics, nor a general awareness of the need for cleanliness, so wounds usually became infected. And because there were no antibiotics the infections could not be treated. Blood transfusions were rarely attempted, and were usually fatal anyway because there was no knowledge of blood types and incompatibility.

Huskisson had lost lots of blood and his chances of surviving without replenishment were faint at best. His extremities were becoming cold, and in the hope of rallying him the physicians administered small quantities of wine and brandy. They also applied hot bricks to his right foot, and to his arms and sides. His pain was excruciating, made all the worse by the constant and violent contractions of the mutilated muscles of his left leg. When the physicians tried plying him with more alcohol some time later, he declined, saying it would only prolong his life, which he was anxious to see at an end. He remained lucid throughout, and requested that a codicil be drafted to his will. This was undertaken by a solicitor named Wainwright, and when it was done Huskisson went to great lengths to sign his name legibly.

When the Duke's sombre procession was within four or five miles of Manchester the crowds had become so dense that they swarmed across the track, forcing the trains to a slow crawl. Parting like a flock of sheep as each train inched its way forward, the 'lower orders' closed in again at the rear as each train went past. The bridges were crammed with people, and while the majority were in a jubilant mood there were some overt displays of dissent. A tricolour flag, reminiscent of revolutionary France, fluttered from one bridge, while a banner with 'Vote by ballot' was waved from another. This protested the current system where votes were registered openly, in public, a practice subject to considerable abuse. As they got closer to Manchester 'the cheerings of the great body of the crowd were interspersed with a continual hissing and hooting from the minority'. There were cries of 'No Corn Laws!' and sundry other exhortations. Some angry weavers pelted the carriages as they went by, but, surprisingly, they did not target the Duke's carriage. In a field near Eccles 'a poor and wretchedly dressed man had his loom close to the road side, and was weaving with all his might . . .'.

At the approach to Manchester the trains had to cross a bridge over the River Irwell. In anticipation of trouble, the bridge was flanked on either side by soldiers of the 35th Regiment. As the Duke's carriage went past they snapped to attention, presenting arms.

The procession had been expected for lunch, but did not arrive until just before four o'clock. As the weary and despondent travellers stepped down from their carriages they were led away to the huge company warehouse, built beside the station to store goods in transit. The upper rooms of the magnificent wooden building had been laid out with 'a most elegant cold collation . . . for more than one thousand persons'. 'The wines were of the most costly and sumptuous description, and the viands plentiful and exquisite.' The Duke and his party declined to leave their carriage, and he spent some considerable time greeting the crowds. Stooping over the handrail of his carriage, he shook hands with 'thousands of persons', and kissed countless small children, lifted up to the obliging gentleman by their mothers.

Meanwhile three of the eight locomotives were detached from their trains and shunted on to the southbound line, the same track as the Duke's train, so they could run into Eccles to take on water. Just why water was not available at Manchester remains unclear. The intention was for these locomotives to return to Manchester and reconnect with their trains.

As more people flooded into the station Mr Lavender, the head of Manchester's constabulary, became increasingly apprehensive about the press of people around Wellington's carriage. At length he 'entreated that the duke's train should move on, or he could not answer for the consequences'. Some of Lavender's concern was doubtless for the safety of the public when the train attempted to head back towards Liverpool. But he may have been more anxious for the well-being of its distinguished passenger – if the radical elements in the crowd changed the mood of the masses, he would be unable to avert a disaster. The order was duly given and *Northumbrian*, hauling the Duke's short train, inched its way through the crowds. It was closely followed by a second locomotive hauling one of the other trains. The other six trains remained where they were.

Once the two trains were clear of Manchester and the crowds had thinned they were able to go faster. But they had not gone very far before they met up with the three locomotives that had gone to Eccles for water. These engines were ahead of *Northumbrian* and could not let it pass because the sidings were temporarily out of use. The only way they could return to Manchester was if *Northumbrian*, and the accompanying train, went too. But running the gauntlet of the crowds was out of the question, so the three locomotives had to travel back to Liverpool with the others. This left six trains in Manchester with only three locomotives to haul them.

When the Duke's vanguard reached Eccles they stopped to enquire after Huskisson. The news was not good 'and this intelligence plunged the whole party into still deeper distress'. Their sombre mood was matched by the weather, and the day that had begun with such bright promise had turned wet and gloomy.

When the three engines failed to return to Manchester those left behind realised something must have gone wrong. Rather than waiting for their unlikely return, the resourceful locomotive men put an alternative plan into action. Coupling all the coaches together, they formed a single train some 300 ft long. The remaining three engines, apparently *Rocket*, *North Star* and *Arrow*, were shunted into position at the head of the column and hitched together.

Pulling in unison, the three locomotives managed to get the enormous train moving, albeit slowly. It was then 5.20 in the afternoon and the rain was falling heavily, adding to the discomfort and difficulties of the crews on the unprotected footplates. But the rain thinned the crowds, and they were soon able to attain their top speed of about 10 mph. Reaching Eccles shortly after six o'clock, they enquired after Huskisson, receiving the same melancholy news as those before them.

Huge crowds had gathered at Liverpool in anticipation of the triumphal return of the procession. Conservatively estimated at 100,000, this was twice the number who had seen the cavalcade leave. Estimates of the total number who witnessed the opening ceremonies ranged from half a million to a million. At first the civil and military forces had managed to keep the crowds off the line, but as the afternoon wore on and the crowds grew, this became impossible. By three o'clock all the streets leading to the railway had become a continuous stream of jostling humanity, all caught up in the excitement of the day and completely unaware of the tragedy that had befallen Huskisson. That the accident had occurred several hours before and only twelve miles away underscores the glacial speed at which news travelled back then.

By five o'clock the procession was long overdue and the impatient throng was becoming unruly. Then a rumour began circulating that a serious accident had occurred. People with loved ones aboard the trains became most distressed and the rest of the crowd became increasingly impatient. By six o'clock patience had worn dangerously thin, but it was not long afterwards that the first locomotive was spotted emerging from Olive Mount cutting. Within a short time of the passengers alighting, news of the tragedy spread through the crowd, causing 'a painful sensation in every bosom'.

The Duke of Wellington was not among the returning passengers. He had alighted at Roby, on the Manchester side of Olive Mount, to drive to the estate of the Marquis and Marchioness of Salisbury. A few miles before stopping for the Duke's party they had come to a siding, which allowed three of the locomotives to return to Manchester to assist the stranded trains.

Having just crossed Sankey viaduct, the long, slow train out of Manchester pulled to a halt at Parkside to take on water. It was getting dark, and as the drivers peered ahead into the gloom they caught the very welcome sight of the three returning engines. Two of them were coupled to the front of the train, while the third was sent ahead to check the way was clear. Using a time-honoured method of railwaymen, the fireman of the lead locomotive lit the way with a burning torch of tarred rope. It was a miserable journey on a foul night, and by the time they reached Liverpool 'only a few drunken fellows' of the 'vast multitude' remained to cheer them. They pulled to a halt at Crown Street station at ten o'clock. Huskisson had died exactly one hour before.

CHAPTER 12

Boom and bust

A railway map of England for the 1830s would have been essentially blank, with just a few spidery traces extending on to the page. These exploratory feelers were mostly centred in the north-east, around Newcastle, and in the north-west, near Liverpool. A dozen years later, though, and the country was cobwebbed with railways, and most major towns had been connected to the system. Not even George Stephenson could have predicted such explosive growth.

Although the Liverpool & Manchester Railway was built to carry goods and passengers, the directors always expected freight haulage would generate most of the revenue, with the travelling public comprising a relatively small part of the business. Given that the railway joined one of the greatest seaports with the premier manufacturing centre in the country, this was a reasonable expectation.

On the day after the grand opening, a special excursion train set off from Liverpool at noon, hauled by *Northumbrian*. The journey to Manchester took just under two hours, including stops. Each of the 130 passengers paid

14*s* (£43) for the return trip, more than many of the lower-paid workers earned in a week. Regular services began the next day, with three train departures from each terminus. The trains carried over 100 passengers, giving a total capacity in excess of 600, but only first-class carriages were available, which probably explains why less than half the seats were filled. The number of passengers steadily grew, though, and there was a marked increase the following week when second-class fares were introduced.

The first-class carriages were similar to those seen on turnpike roads, with padded seats and springs. Like a stage-coach, the roof was piled with luggage, and many of the trains posted guards atop the first and last carriages to sound a bugle when they were about to set off. Second-class passengers had to make do with hard seats and open carriages, though there was usually some sort of roof. Although their accommodation was a little spartan, the 4*s* (£12) fare for a one-way trip brought them greater comfort and a shorter travelling time than going by road, and was also much cheaper. For comparison, the thirty-five-mile journey by stagecoach took four and a half hours, and although the operators lowered the fares when the railway opened, it still cost 10*s* (£30) to ride inside, and 5*s* (£15) on top.

By the end of the month passenger traffic had tripled, and additional trains had to be scheduled to meet the demand. The original eight locomotives would soon be stretched to the limit. The directors had ordered four more at their 20 September meeting, and by the following spring they would need two more. Such explosive growth in passenger traffic had caught everyone in the company off guard, which may explain why freight traffic lagged behind, and why it was not until the beginning of December that the first goods train was despatched. Hauled by *Planet*, the eighteen-wagon train weighed eighty tons and completed the journey from Liverpool to Manchester in just under three hours. The cargo included 100 bales of cotton, 200 barrels of flour, thirty-four sacks of malt and sixty-three bags of oatmeal.

Road haulage co-existed with railways for longer than did stage coaches.

By the year's end, after less than four months in operation, the company had made a profit of £14,000 (£854,000), allowing it to pay the shareholders a modest dividend of 2 per cent. Such an outstanding performance is all the more remarkable considering this was the first railway of its kind, and almost every aspect of its operation, from ticketing passengers to handling freight, had to be developed *de novo*. The next set of figures was even better and the returns for the first five months of the following year showed a profit in excess of £30,000 (£1.8 million), allowing a dividend of 4.5 per cent. Such a promising start captured the attention of investors throughout the country, and the volume of business just kept on growing. During 1831, the first full year in operation, the railway carried 445,047 passengers and 43,070 tons of freight. By 1835 the passenger traffic had increased by 6.5 per cent and the tonnage of goods had leaped more than fivefold to 230,629 tons. These numbers did not include coal traffic, which saw a tenfold increase in the same period, from 11,285 tons to 116,246 tons.

If the rest of the country had not yet realised the revolutionary changes the railways would cause, the operators of the stagecoaches that plied between Liverpool and Manchester were in no doubt. Twenty-six coaches served the route before the railway opened, but these had been reduced to a dozen by the end of 1830. Stagecoaches could never compete with railways, and an entire way of life was about to come to an end, affecting everybody and everything associated with the business, from turnpike roads and coaching inns to wheelwrights and farriers. Within a few decades coaching would be nothing more than a nostalgic memory, committed to the writings of Charles Dickens, and to the Christmas cards of later years. No more yarn-spinning drivers in their topcoats and hats. No trumpeting horns, no flying hoofs, and no passengers hanging on for dear life. Gone too the jovial landlords, buxom cooks, steaming punch and roaring fires. But such fond recollections overlooked the grim reality of travelling by coach, with gruelling hours spent on rutted roads being shaken senseless. Footpads and highwaymen notwithstanding, road travel could be downright dangerous too, as George Stephenson recalled in a letter to Longridge, written in 1835:

> I left London last night and arrived here this morning, without the repetition of the upset I had in going up by 'Hope' which I daresay you would see by the papers. I *saw* she was *going to upset* and being inside made use of a little science which brought me off safe. I never saw such a sight before, passengers like dead pigs in every direction and the road a sheet of blood. Two, I apprehend, will die.

The rapid rate at which the railways spread is quite remarkable. When the Liverpool & Manchester Railway opened in 1830 only about 100 miles of steam-operated track had been laid. But a decade later this figure had jumped to about 1,400 miles, owned and operated by some forty independent companies. In the early days, between 1831 and 1834, Parliament sanctioned only four or five new railways each year, but the number steadily increased, and by 1844 it was up to about thirty. This increased to almost 100 the following year, and then to over 200 in 1846. Railway construction at the time was providing a quarter of a million jobs, representing approximately 4 per cent of the male working population. Irish immigrants made up the greater portion of this work force, together with itinerant farm labourers from the rural south of England. The average prices of railway shares followed a similar trend, increasing steadily from about £60 to £70 between 1830 and 1835, then proceeding in a series of bounds to peak at about £150 (£9,600) in 1845. Railway stocks dominated the capital market and by 1843 the seventy railway companies that were quoted on the London stock exchange had an authorised capital of £57 million (£3,876 million).

Those with a vested interest in the old ways of transport loathed the 'infernal railroads', as did many landowners, but investors and speculators just loved them. And the love affair, which blossomed with the opening of the Liverpool & Manchester Railway, soon grew into a torrid, all-consuming obsession, aptly named 'Railway Mania', which peaked during the mid-1840s.

Not all proposed railway schemes were completed, and of those that were finished not all were financially successful. Some schemes were not even genuine, but there was such heightened expectation of the new technology that investors continued spending their money, regardless of any bad news. Such a heated market was ripe for fraud and there was no shortage of scoundrels ready to take advantage of the feeding frenzy. The biggest of all the rogues was George Hudson, the man they called the 'Railway King.'

Hudson (1800–71) was a farmer's son and after leaving the local school he became apprenticed to a draper in the city of York. His employer was obviously impressed and eventually made him his business partner. But it was in business, rather than in drapery, that Hudson's true talent lay. He might have remained in the textile trade, where he would no doubt have made a good living, but when he was twenty-seven a great-uncle died, leaving him an enormous legacy of £30,000 (£1.7 million). Many young men in a similar situation would have been happy to bank the money and live a life of leisure, but Hudson wanted to make even more money. He realised the best way of doing so was to speculate on the stock market.

The Liverpool & Manchester Railway was still under construction, but Hudson, like George Stephenson, could visualise a day when the entire country would be covered by a network of railways. He passionately wanted to see this happen for the good for the nation – it would also be good for George Hudson – so he put his fortune to work. He bought up shares, set up companies, traded, manipulated, speculated, misappropriated and generally pumped up the value of railway stocks. Recognising that so many independent companies, each operating only a few miles of line, could not survive, he devoted much of his energy to amalgamating them. This phase of consolidation benefited the growing railway network, and increased the size of Hudson's empire: by the mid-1840s he controlled almost one-third of the 5,000 miles of existing track.

George Stephenson met Hudson in 1834, and the two men took an immediate liking to each other. Stephenson worked on several of his railway projects, and Hudson exploited their relationship to attract investors to his schemes. But the canny engineer began to see Hudson for what he really was, and eventually refused to have anything more to do with him. 'Hudson has become too great a man for me now . . .' he wrote to Longridge towards the end of 1845. 'I have made Hudson a rich man but he will very soon care for nobody except he can get money by them . . .'

Hudson was politically active, serving for many years as leader of York's Conservative Party and holding a succession of municipal positions, starting as a city councillor in 1835, alderman the following year, and Lord Mayor in 1837. He successfully stood for office as the candidate for Sunderland in the general election of 1845, and served a four-year term. At the time of his election his railway holdings were in excess of £300,000 (£19.3 million). Railway mania was then running at fever pitch, as captured in some lines of verse in the *Illustrated London News*:

Railway Shares! Railway Shares!
Hunted by Stags and Bulls and Bears
Hunted by women – hunted by men

Speaking and writing – voice and pen
Claiming and coaxing – prayers and snares
All the world mad about Railway Shares!

All this sat very well with Hudson, but others did not like the way things were going. Isambard Kingdom Brunel (1806–59), one of the greatest engineers of the nineteenth century, wrote to a friend that 'the railway world is mad. I am really sick of hearing proposals made. I wish it were at an end.' George Stephenson was pessimistic and predicted that 'thousands of people will be ruined . . .'. Brunel got his wish and Stephenson's prediction came true, because the over-inflated bubble burst and share prices plummeted. By 1850 shares had fallen to half their peak values of 1845. Hudson came crashing down and in the ensuing investigation the full extent of his wrongdoing was revealed to a scandalised public.

CHAPTER 13

Beginnings and endings

A boy, some twelve years of age, sits at a school desk studying in silence. The trees outside have already begun to turn and Bodmin Moor in the distance lies resplendent in the russet tones of autumn. Just moments before, the master, seated imperiously at the front, had asked the boy whether he had received any particular news from home. Puzzled by the question, the boy had answered in the negative.

The curiosity of the other boys has scarcely died down when a tall man enters the room. He wears a broad-brimmed hat of fine plaited straw which is as foreign to England as his mahogany tan. Glancing around for a moment, the stranger strides purposefully towards the master's desk at the far end. Half-way there he pauses, removes his hat, and asks if his son Francis is present. A hushed expectancy falls upon the room and all eyes fix upon the stranger. The boy who had been at the centre of the other boys' stares feels his heart pounding fit to burst. Could this sun-worn man with the clear blue eyes be his father? Was this the same man who had carried him down the quayside steps so he could wave him goodbye? The man who had sailed away and out of his life for eleven years?

Richard Trevithick spent the next six months getting to know his son Francis. Each day they would sit together, the one working on some new scheme or invention, the other observing and learning. On one occasion they were engrossed in the relation between the size and weight and speed of a swallow they had shot. The inventor hoped to use the information to calculate what size wings might be needed to carry a man's weight. Cube roots of measurements kept recurring in these calculations and young Francis attempted to find an explanation by reciting passages from his school books. But this did not help his father's comprehension of the problem.

Trevithick's interest in man-made flight was interrupted a month later when he received word from Gerard, a South American acquaintance, who had just returned to England. Gerard was planning to interest London speculators in a new mining scheme in Costa Rica, and Trevithick was keen to participate. Meanwhile, Trevithick was working on two projects which he hoped would be of interest to the Admiralty. The first was a mechanical gun-carriage that would enable a cannon to be operated by only one seaman. Driven by the gun's recoil, he calculated, the device would increase the rate of firing fivefold over conventional ordnance. The second project was an iron ship with an improved steam engine.

Early in 1828 his old acquaintance, Lord Cochrane, returned to England. Trevithick was delighted because he was sure a man of his background and influence could put in a good word for him with the Admiralty. But, regardless of any assistance Cochrane may have rendered, the Admiralty summarily rejected Trevithick's gun carriage as 'wholly inapplicable to practical purposes'. Nothing became of his iron ship, either. Well acquainted with rejection, the Cornishman carried on with his other projects without breaking stride.

One was an appeal to Parliament for compensation for the fuel savings his high-pressure steam engines had made since their introduction. He estimated the savings in Cornwall alone to be in excess of £500,000 (about £25 million). Transmitting the petition through his old friend Davies Gilbert, now a Member of Parliament, he moved close to London so he could lobby MPs, and also monitor the Costa Rican mining venture. But it was of no avail and he never received a penny in compensation. His friend Gerard was having no better success with potential mining speculators in the City. After trying for one or two more years with the same results, Gerard decided to try his luck on the Continent. He failed there too, and died in Paris in poverty. But the dream of making a fortune from the riches of the New World lived on with his friend. Attending a meeting with some prospective investors, Trevithick beguiled them with stories of the fabulous mineral wealth of the Andes. Several of the gentlemen were so impressed they wrote him a cheque for £8,000 (about £470,000) for his copper concession in Peru but he rejected their offer out of hand, believing his holdings to be worth vastly more. Needless to say he made no money whatsoever from the mining interests.

Trevithick was not interested in politics, but the passage of the Reform Bill in 1832 inspired a new project. This landmark legislation abolished rotten boroughs, replacing them with new ones more representative of the population. It also extended voting rights, though voters still represented only about one working man in ten. The Reform Act was surely worth commemorating with a monument, and Trevithick's idea was to build a column, 1,000 ft tall, made of gilded cast iron plates. Public access to the top of the edifice would be provided by a 10 ft diameter floating piston, fitted snugly inside a central tube and driven by compressed air. His proposal appears to have generated considerable interest, receiving the approbation of some of the most influential men in the land, including six peers of the realm.

On 1 March 1833, nine months after presenting his proposal, he received a letter from His Majesty's secretary informing him the design had been submitted to the King. This was most promising news and Trevithick knew he was on the brink of securing the contract of a lifetime. Three weeks later he was dead.

Trevithick died in poverty in a rented room at the Bull inn, Dartford, a Kentish village some fifteen miles from the City of London. There were no family members by his side, and they learned of his death only through a note written by an acquaintance. His funeral was paid for by his friends and acquaintances, apparently assisted by the sale of a gold watch he had brought back from South America. They also protected his grave from body-snatchers, who were prevalent in the neighbourhood at the time.

Trevithick would have died a rich man had he not spent his life in quest of the unattainable. But, in spite of all the setbacks and disappointments in life, most of them self-inflicted, his spirit remained remarkably buoyant, right to the end. Writing to Davies Gilbert, a few months before his death, he said:

> I have been branded with folly and madness for attempting what the world calls impossibilities, and even

> from the great engineer . . . James Watt, who said . . . that I deserved hanging for bringing into use the high-pressure engine. This so far has been my reward from the public; but should this be all, I shall be satisfied by the great secret pleasure and laudable pride that I feel . . . [for] bringing forward and maturing new principles . . . of boundless value to my country. However much I may be straitened in pecuniary circumstances, the great honour of being a useful subject can never be taken from me, which to me far exceeds riches.[3]

A year after Trevithick's death his old acquaintance, Lord Cochrane, Earl of Dundonald, approached the Liverpool & Manchester Railway with a request. He wanted to use *Rocket* as a testbed for his newly invented rotary steam engine. He was sure the expense of fitting the engine to the locomotive would not exceed £30 (£1900), and agreed to pay any additional expenses. The company approved, and went ahead with the conversion, but the costs were soon nearly triple his lordship's estimate, and the work was still not complete. The trial eventually took place around 22 October 1834, but was a complete failure. Dundonald thanked the company all the same, and asked for a statement of expenses. His revolutionary new steam engine never did work, and it is unclear whether he ever settled his bill.

The Stephensons, in contrast to Trevithick, had made a good living from their enterprises, and both would depart the world rich men. The success of the Liverpool & Manchester Railway had opened the nation's eyes to the enormous possibilities of railways, and a number of new lines were proposed. George Stephenson, the foremost railway builder in the land, was much in demand, and if his name appeared on the prospectus as the engineer of a new railway project a full list of subscribers was almost guaranteed. Robert, meanwhile, continued developing better locomotives, ably assisted by William Hutchinson and the rest of his skilled team at Forth Street. Improvements came so fast that the locomotives used during the opening ceremony of the Liverpool & Manchester Railway were all superseded within weeks. *Planet*, the first to surpass them, lent its name to a whole new class of locomotives which the company built for the next several years. Its boiler was the same size as *Northumbrian*'s, so were its cylinders, but these were placed in a novel position. Instead of being mounted external to the wheels, as in *Rocket* and its descendants, the cylinders were placed internally, beneath the smokebox. The pistons were connected to the rear driving wheels by a cranked axle. Power transmission was smooth, and placing the cylinders beneath the hot smokebox reduced heat losses considerably, improving efficiency and therefore fuel economy.

In later life Robert would look back upon the time developing locomotives in Newcastle as the happiest years of his life. But, like his father, he became increasingly interested in building railways rather than locomotives and began developing a career as a civil engineer. Unlike his father, though, he joined the Institution of Civil Engineers, that body of predominantly London engineers whom George held in such contempt.

The first major railway project after the Liverpool & Manchester line was the London & Birmingham Railway. This colossal undertaking would form the first link between the industrial north and the capital. Robert was engaged as the engineer, assisted by Thomas Gooch, who had begun his railway career working for George Stephenson as a draughtsman on the Liverpool & Manchester Railway.

Robert ran into fierce opposition during the survey, just as his father had done for the Liverpool & Manchester line. One of the major opponents was the Grand Junction Canal – as operators of the main artery between London and the Midlands it felt justifiably threatened by the proposed railway. There was determined opposition from the turnpike roads too, with dire warning of the perils of travelling by train. Pamphlets were distributed, warning the public to 'beware of the bubbles', implying that the new railway was just another stock promotion scheme that would burst like all the others. The railway promoters were described as quacks who should be confined to Bedlam, the infamous London mental hospital. A rumour was even spread that locomotives had been found to be so useless and downright dangerous on the Liverpool & Manchester line that they were to be replaced by horses. The directors of the company were sufficiently disturbed by the rumour to issue a public denial.

Landowners also fought to stop the railway from crossing their land, the strength of their opposition being out of all proportion to the relatively small amounts of their large estates that would be affected. Sometimes hostilities became so intense that the surveyors were forced to take their measurements at night, using dark lanterns. One clergyman was so antagonistic that the crew conducted their survey on a Sunday while he was busy saving souls. Sir Astley Cooper, the eminent London surgeon, received Robert Stephenson and his crew most civilly when they called at his country estate in Hertfordshire. He listened attentively to all of Stephenson's arguments in favour of railways, but remained adamant that none would cross his land:

> Your scheme is preposterous in the extreme [said the courtly old gentleman], positively absurd . . . You are proposing to cut up our estates in all directions for the purpose of making an unnecessary road . . . Why, gentlemen, if this sort of thing be permitted to go on, you will in a very few years *destroy the noblesse*!

It seems that Sir Astley Cooper perceived the railways' march of progress as a real threat to the privileged ruling class. As they left the house Stephenson commented to the others that it was 'really provoking to find one who has been made a "Sir" for cutting that wen out of George the Fourth's neck, charging us with contemplating the destruction of the *noblesse* . . .'.

Regardless of all the opposition and difficulties, the survey was completed, and the new railway Bill went before a Commons committee in 1832. And, like the battle royal that had marked the first attempt to pass the Liverpool & Manchester Railway Bill, the opponents used every device, fair and foul, to defeat its passage. This time their target was Robert Stephenson. His father had been outgunned and outmanoeuvred, but the opposition found Robert a most formidable opponent. Their disparaging remarks about his youth – he was still only twenty-nine – had little effect. They tried probing him on the minutiae of the survey, hoping to catch him out the way his father had been exposed during his parliamentary ordeal. George was hopeless with details, but his son was meticulous in all aspects of surveying. At one point he was attacked for the steep angle of his proposed cutting through the chalk at Tring, northwest of London. The opponents' counsel argued that such steep slopes would inevitably collapse upon the line. Stephenson disagreed. When the proceedings ended for the day he recalled that the eminent engineer, Thomas Telford, had used a similarly steep angle when cutting a road through the same chalkbed at Dunstable, not ten miles from Tring. Although Stephenson had been under cross-examination for the previous three days, and had worked through each night without proper sleep, he was determined to visit Dunstable before the next day's session began. Working in his room until midnight, he took a quick meal, had a short nap, then hired a fast carriage to take him on the forty-mile journey, accompanied by Thomas Gooch. Arriving in Dunstable at dawn, they were delighted to confirm that Telford had used the same steep angle as he planned using. They returned to London knowing they could now demolish the oppositions' case against their chalk cutting.

The Bill passed in the Commons by a large majority, only to be defeated in the upper chamber of the House of Lords the following month. Robert was crestfallen, but Lord Wharncliffe, the chairman of the Lords committee, was reassuring. He attributed the failure entirely to the landowners in the Upper House and not to any shortcomings in the proposal itself. What was more, Stephenson's outstanding performance under cross-examination had impressed a good many people, assuring his career as an eminent civil engineer. And, as Lord Wharncliffe pointed out, there were ways of getting their lordships to approve the resubmitted railway Bill during the next session.

When the Bill was reintroduced during the 1833 session it passed through both Houses with little opposition. The reason for its safe passage through the Lords on the second occasion was largely because the company had nearly trebled its land valuations from the original estimate of £750,000 (£47 million): their lordships had simply wanted more money for their land. Robert was appointed engineer-in-chief of the new railway, and he and his wife left their home in the north to take up residence in London.

That same year another major project obtained parliamentary approval, the Grand Junction Railway connecting the Liverpool & Manchester Railway with Birmingham, thereby linking the capital with the north. The line was surveyed by Joseph Locke, ostensibly under George Stephenson's supervision, though Locke, the most talented of all his assistants, was more than capable of completing the task unassisted. Indeed, he was far more thorough than George Stephenson, and made an excellent job of the survey. Although the two men had worked together for many years, relations between them had become strained, and Locke knew it was time for him to strike out on his own.

Locke's completion of the survey fulfilled his obligation to Stephenson and he felt at liberty to apply for the position of chief engineer. The company had been so pleased with his work that it was very receptive to the idea, but Stephenson was predictably aggrieved. The ensuing dispute became so unpleasant that Locke withdrew his candidacy, but the company did not want to lose either of them, so it proposed a compromise. Locke would take charge of the northern half and Stephenson the southern sector.

Locke was meticulous, tendering contracts for specific segments of the line, none of them exceeding ten miles in length, and prospective contractors were supplied with clear specifications and requirements. Stephenson, in contrast, muddled along the way he always did, with inaccurate measurements and rough drawings, leaving contractors scratching their heads. As a result, Stephenson had barely got his first contract signed when Locke obtained his last signature. The company could not let such a situation continue, but did not want to offend Stephenson, so it suggested the two men became joint engineers for the entire project. Locke, knowing his old mentor the way he did, recognised the futility of such an arrangement. There was only one solution to the problem and Stephenson, mellowing in his middle years, eventually withdrew, leaving Locke as the chief engineer of the entire project. But it was not a gracious withdrawal, and George Stephenson never forgot or forgave the blow to his pride. From that time on Locke became *persona non grata.* Robert Stephenson and Locke had been good friends for years, but, out of loyalty to his father, Robert ended their friendship.

Once Locke took charge of the entire line, work proceeded at a pace, and the Grand Junction Railway opened just two years later, in 1837. There was none of the pomp and ceremony that had marked some of the previous openings, and there never would be another lavish spectacle like the inauguration of the Liverpool & Manchester Railway. This was only to be expected, because almost two dozen new railways had been opened since that occasion.

The year 1835 was one of the busiest of George Stephenson's life. During that, and the following two years, he travelled some 20,000 miles by post-chaise, looking over prospective railway routes throughout the Midlands and the north. Being able to pick and choose projects, he was spending most of his time as a consultant rather than as a principal engineer. Released from the responsibility of running a major engineering project, he could now do what he liked and did the best, which was making preliminary surveys of new lines. He had an outstanding talent for divining the best route to take, knowing which features to avoid and which ones to face. He was also adept at assessing the potential utility of a proposed line and in exploiting local resources. If, for example, there were coal deposits near by, he might decide to make a detour, mutually exploiting the railway and the resource.

George, now in his mid-fifties, had acquired both public recognition and great personal wealth, and no longer felt threatened by his detractors. Nor was he so irrationally jealous of his rivals. He could afford to lower his guard and become less competitive, more agreeable – the likeable person those close to him knew he always was. He was more than content to hand over all the major engineering projects to Robert.

The London & Birmingham Railway occupied Robert Stephenson's energies from 1833 until its completion in 1838. Employing upward of 20,000 men, and never fewer than 12,000, it was the largest engineering work hitherto undertaken in Britain. The 112-mile line was just over three times longer than the Liverpool & Manchester Railway and presented Stephenson with engineering challenges that dwarfed even the Sankey viaduct and Olive Mount. These included the mile-long viaduct across the River Avon, the chalk cutting at Tring, over two miles long, and an embankment, just to the north, that was almost six miles long. But the greatest challenge of all lay in wait for him in the sleepy village of Kilsby, four miles from Rugby.

Thomas Arnold, the famous headmaster of Rugby School, was familiar with the countryside around Rugby, and when Stephenson happened to call upon him during his survey of the area he was pleased to share his knowedge. When the engineer told him of his intention to tunnel through the nearby hills, Arnold made the prophetic statement

'I confess I shall be much surprised if they do not give you some trouble.'

If the 1.4 -mile tunnel had been cut through hard rock the job of excavating, though challenging, would have been straightforward. But the workmen soon encountered quicksand, millions of gallons of the wet slurry, through which it was impossible to tunnel. The contractor in charge of the project was unable to cope with the quicksand, or with the stress of the job, and soon fell ill. He eventually died, leaving Stephenson to battle with the horrendous problem for the next two years. Kilsby tunnel was Robert Stephenson's Chat Moss, but on a monstrous scale that took its toll of human lives as well as of resources. Most of the deaths occurred through cave-ins, and many an unfortunate navvy was buried alive when his shovel inadvertently cut through a seam of the treacherous fluid. Upward of 1,200 men and 200 horses were employed at the peak of the operation, and the original cost estimate tripled to £291,000 (£17.8 million). It was only after thirteen steam engines, each pumping at the rate of 1,800 gallons a minute, had laboured for nineteen months that the battle was finally won. The tunnel was a major contributor to the cost and time overruns of the entire project, which finally reached £5.5 million (£326 million), just over twice the original estimate.

When the end of the tunnelling was finally in sight the entire engineering staff convened at the Dun Cow, a local hostelry, for a night's revelry. Father and son were in excellent form and when Robert took his leave at 2.00 a.m. his father was still going strong. Jolly old George continued for at least another two hours, and the most dedicated souls kept the party going until breakfast.

Completing the railway was a herculean task that needed an engineer of Robert Stephenson's outstanding ability. He could command an army of workmen like a battlefield general, overseeing and co-ordinating every part of the plan and every detachment of men. He was self-confidence personified to everyone around him, but underneath the surface he was plagued with self-doubt. He once confided in a friend that 'I sometimes feel very uneasy about my position. My courage at times almost fails me and I fear that some fine morning my reputation may break under me like an eggshell.'

There were two opening ceremonies, both fairly restrained. During the first, held on the summer solstice of 1838, a gathering of directors, engineers and workmen assembled beneath one of the ventilation shafts in Kilsby tunnel to witness the last brick being ceremonially tapped into place with a silver trowel. When the band had finished playing 'God save the Queen' – Victoria had been on the throne for a year and a day – the company marched to the north end of the tunnel to enjoy an open-air feast. On the official opening, three months later, a special train conveyed the London directors and their guests to Birmingham, where they joined their counterparts for a banquet at the Royal Hotel.

Linking London by rail had consequences reaching far beyond the obvious one of improving access for people and goods between the centre of commerce and the industrial north. At that time London had a population of under 1.5 million and occupied an area a fraction of its present size. It was surrounded by farms and open countryside, and the site of Euston Station, the railway's terminus, was a pasture where a herd of a thousand cows grazed, supplying milk for the city. Regardless of how many people flocked into London to earn a living, the physical growth of the city was in large part restricted by the need for arable land to supply the population with fresh food. Perishables like milk could not be brought in from afar because roads and canals were just too slow, though meat could be brought into the city on the hoof. But the railway changed everything. Farms and fields gave way to bricks and mortar, and the sight of herds of livestock being driven through the streets of London by smocked rustics became a thing of the past.

It had been expected that the first locomotives for the London & Birmingham Railway would be supplied by the Forth Street Works, builders of the finest engines in the world. But Stephensons' detractors caused such uproar with their cries of 'foul' and 'monopoly' that the railway was obliged to pass a resolution prohibiting further purchases. Robert was distressed by all the unpleasantness, and was also concerned for the future of the locomotive company. But his concern was entirely unfounded and the Forth Street order book just kept on filling. Aside from strong domestic sales there were exports to Europe, North America and Russia. The flame his father had kindled in Killingworth all those years before was spreading like wildfire, changing the tempo and mode of life for ever.

The claim that the Stephensons had a virtual monopoly of railway construction was certainly true during the early years – if George Stephenson could have had his way it would have remained that way for ever. But by the mid-1830s other players had entered the field, including two of the other Rainhill contestants, and some of George's former assistants, including Joseph Locke and Charles Vignoles. Locke would rise to such eminence as a civil engineer that when he died *The Times* counted him as one of a triumvirate of great engineers, along with Robert Stephenson and Isambard Kingdom Brunel.

Brunel, who was also the son of a distinguished engineer,

had many of Robert Stephenson's characteristics. He was young and bright, charismatic, fiercely independent, doggedly determined, and had an insatiable appetite for hard work. Add to this his confidence, a bulldozer approach to obstructions, and an unwillingness to suffer fools gladly, and a picture emerges of the inexorable force that was Brunel. Robert, who was only three years his senior, had the highest regard for his engineering abilities. The feeling was mutual and the two men became firm friends, always there to support the other.

Brunel built bridges, railways and transatlantic steamships. His *Great Western* was the first steam-driven ship on regular transatlantic passenger service, and *Great Eastern* was the largest ship built during the nineteenth century. Born in Portsmouth, on the south coast, he concentrated his engineering activities in the southern part of the country, which suited George Stephenson very well. His first railway was the Great Western, linking London with the seaport of Bristol. Of all his projects this is the one for which he is most remembered. He was appointed chief engineer in 1833, while still a month away from his twenty-seventh birthday, and the project received parliamentary sanction two years later.

As an engineering achievement, the Great Western Railway probably surpassed the London & Birmingham Railway, which it exceeded in length by just five miles. The line incorporated some of the most outstanding examples of railway architecture of the era, including the brick bridge across the Thames at Maidenhead and the imposing Box tunnel. Maidenhead bridge is remarkable for the extreme flatness and breadth of its two main arches, and was, at the time, the largest span that had ever been attempted in brick. The calculations Brunel made in his notebook show

Isambard Kingdom Brunel photographed beside the launching chains of his last ship, Great Eastern.

he had a sound understanding of arch design, but his critics were sceptical of its success. During construction the bricks on the underside of the arches were held in place by wooden supports while the mortar dried. When these were removed it was discovered that some of the lower courses of bricks had separated from the rest. His critics were delighted. However, the contractor, admitting liability for removing the supports before the mortar had set, made the necessary repairs, and the bridge has been standing ever since.

Box tunnel is almost two miles long and was the longest one of its day. Half a mile longer than Kilsby tunnel, it is also much higher, dwarfing the trains that pass through its impressive portal. Brunel built it ramrod-straight, and it is said that on his birthday, 9 April, and on no other day of the year, the rays of the rising sun shine through its entire length. There may be some truth in the legend, because, from the alignment of the tunnel and the position of the sun, it has been shown on theoretical grounds that the phenomenon could occur on some April mornings, though not on 9 April. Direct observations made during the 1960s and 1970s showed that the phenomenon variously occurred on 15, 16, and 17 April, and there is one independent claim that in 1981 it occurred on 9 April. Regardless of the validity of that claim, the author of the theoretical study concluded that Brunel had not deliberately attempted to align the tunnel to produce the solar effect.

When Brunel began work on the new railway he had the advantage of being able to draw upon the experience of a decade of railway construction, primarily the result of George Stephenson's work. By then certain features had become standardised, including the multi-tubular boiler, blast pipe, horizontal cylinders and a rail gauge of 4 ft 8½ in., all of which persisted to fairly recent times. The last feature arose for no other reason than it happened to be the width of the Killingworth wagonway where Stephenson built his first locomotive. There was nothing particular about this gauge, other than its being a convenient width for horses, and all horse-drawn wagonways were of similar width. When Brunel came to build his railway he decided, with sound mechanical reasoning, that a wider gauge would give greater stability, especially at the high speeds he planned for his trains. Not a man to do things by half measures any more than he was to be ruled by precedent, he adopted a 7 ft gauge. The broad-gauge locomotives were characterised by their massive driving wheels, up to 10 ft in diameter, designed for speed. His first locomotives did not perform well, but that changed when he started having them built by Robert Stephenson & Co. The first of these, *North Star*, was delivered in 1837, and it was so successful that eleven more of the same class were delivered during the next five years. Daniel Gooch, the younger brother of Thomas, was the superintendent of the Great Western Railway, and, like Hackworth, he began designing locomotives. These were built by various manufacturers, including Stephenson, but by 1846 the company was building locomotives at its own works in Swindon.

Brunel went on to build other broad-gauge railways, all in the south-west, so there were no incompatibilities with the standard-gauge lines in the rest of the country. But as both systems continued to expand they inevitably came into contact, causing considerable inconvenience. Something had to be done to resolve the problem, and the government appointed the Gauge Commission to review the situation. Brunel, confident of the ultimate success of his superior broad-gauge railway, threw down the gauntlet to put the two systems to the test. A number of trials took place early in 1846, under the watchful eye of the Gauge Commission. Robert Stephenson chose two of his latest long-boiler goods locomotives to champion the standard gauge. One was borrowed from the North Midland Railway, and the other, known as the 'A' engine, was from his own works. Brunel's champion was *Ixion*, built in 1841 and named for the mythical Greek king who was punished by being bound to an eternally revolving wheel. The long-boiler class of locomotives were designed for hauling heavy loads, and while they were very successful in that role, they were unstable at high speeds. During one of the experiments the borrowed Stephenson engine 'jumped off the rails, falling upon her side' while travelling at 48 mph,[,] but *Ixion*, like other broad-gauge engines, was very stable.

Overall, the broad gauge was superior, but it comprised such a small percentage of the total mileage of rail that the commission recommended in favour of the narrower system. This was legislated the following year by the passage of the Gauge Act which made 4ft 8½ in. the standard gauge. Brunel's broad-gauge railways continued operating, but were eventually converted to the standard gauge. The conversion was expensive and protracted, and it was not until 1892 that the last broad-gauge trains stopped running on the Great Western Railway. Remarkably, not a single one of these magnificent locomotives survived in the entire country.

When railway shares were hitting their peak in 1845 George Stephenson, now sixty-four, had already retired and was living at Tapton House, Chesterfield, which he had leased two years earlier. Tapton House stood on a hill overlooking the North Midland Railway, on the edge of Derbyshire's Peak District. He chose the spot because there were coal and limestone deposits in the vicinity, which were now worth exploiting because of their close proximity to the railway. He divided his retirement years between over-seeing the quarrying and mining operations, and pursuing his old love of horticulture. His wealth enabled him to indulge himself, and he built a number of hothouses where he could grow tropical fruits and compete with other gardeners. But he did not vegetate at Tapton House, and continued travelling, visiting friends and acquaintances and meeting new people. By this time almost all the railway companies with which he had been associated had issued him free passes, and he was too frugal not to use them.

George was comfortable with himself the way he was and had several times declined a knighthood, just as he had refused to be nominated to the Royal Society. He was equally at home making the rounds of his old cronies at Killingworth and spending weekends at Drayton Park, the country home of Sir Robert Peel, the Prime Minister. Peel used to invite several celebrated guests to join him during these weekends, and on one particular occasion Stephenson got into discussion with Dr William Buckland on the subject of the formation of coal. Buckland, the discoverer of the world's first dinosaur, was Professor of Geology at Oxford University and was therefore expected to know about such things. The party had just returned from church and were standing on the terrace, watching a train in the distance, trailing a long cloud of steam.

> 'Now, Buckland,' said Mr Stephenson. 'I have a poser for you: can you tell me what is the power that is driving that train?'
>
> 'Well,' said the doctor, 'I suppose it is one of your big engines?'

> 'But what drives the engine?'
>
> 'Oh, very likely a canny Newcastle driver.'
>
> 'What do you say to the light of the sun?'
>
> 'How can that be?' asked the doctor.
>
> 'It is nothing else,' said the engineer: 'it is light bottled up in the earth for tens of thousands of years; light absorbed by plants . . . during the process of their growth . . . And now, after being buried in the earth for long ages in fields of coal, that latent light is again brought forth and liberated, and made to work, as in that locomotive, for great human purposes.'
>
> Like a flash of light, the saying illuminated in an instant an entire field of science.

George Stephenson's idyllic life suffered a major blow on 3 August 1845 with the death of Elizabeth, his second wife of twenty-five years. And it was towards the end of that year that he had his first encounter with serious illness. He was visiting Spain, accompanied by a Liverpool financier, Sir Joshua Walmsley, to consult over a proposed railway line. The trip involved a good deal of travelling across rough country, working long days and spending cheerless nights on the floor of some miserable hovel. The hard conditions seemed to take their toll, and by the time they reached Paris, on the return leg of the journey, he was feeling quite ill. Directly they boarded their ship at Le Havre he took to his cabin with a severe attack of pleurisy. A doctor was summoned, 'who took 20 oz blood' from him.

Blood-letting was one of the standard 'treatments' of the time, the belief being that removing blood also removed poisons and impurities from the body. The patient would be bled either with a sharp instrument or with leeches. Another equally useless procedure was cupping, where a cup was heated in a flame then held tightly against the skin. As the remaining air in the cup cooled, it formed a partial vacuum, drawing blood to the surface of the skin.

Writing to Longridge about his ordeal, he says that when they arrived in London they sought the 'best advise; they got two Physicians to me; they put me to bed and cupped me on the right side; how much blood they took by cupping I cannot tell . . .'. In spite of the attentions of the medical profession, he made a full recovery, and life at Tapton House returned to normal.

Stephenson did not remain a widower for very long, and early in 1848 he married his housekeeper. Six months later he was struck by a second attack of pleurisy. The fever lasted for about ten days and at one point he seemed to be recovering. But this was followed by a relapse, and on the morning of 12 August he suffered a sudden haemorrhaging from the lungs. He died at noon that same day, aged sixty-seven. He was buried in the local churchyard and the townsfolk of Chesterfield showed their respect by closing their shops and businesses and joining the funeral procession. They were accompanied by many of his former workmen, all of whom remembered him as a good and kindly employer.

Winning at Rainhill was the only thing that mattered at the time of the trials, but one of the contestants, reviewing his life towards its end, maintained that losing was the most fortunate thing that could have happened to him. 'With success would have come immediate prosperity and corresponding temptation . . .' Ericsson wrote. Instead he had to struggle throughout his life, and in doing so he attained some truly outstanding engineering accomplishments. Among these was a depth gauge that allowed accurate soundings to be taken while a ship was still under way. Prior to this, vessels had to be stationary while a weight was dropped overboard at the end of a calibrated line. It was assumed that the line ran vertically to the bottom, but, because of currents, this was seldom true, and soundings were usually inaccurate. Ericsson's device used the relationship between water depth and pressure and was therefore independent of such errors. Thousands of the devices were sold, making an enormous contribution to the accuracy of maritime charts, saving ships as well as lives.

In 1835 he designed a ship's propeller, and the following year had it fitted to a steamship. Purposely built for the experiment, it was named *Ogden*, after the US consul who helped finance the project. Ericsson invited the lords of the Admiralty to a demonstration and they were accompanied by Sir William Symonds, the Surveyor of the Navy, and other naval and scientific gentlemen. The demonstration was a complete success and *Ogden* towed the Admiralty barge down the Thames with ease, at a steady 10 mph. This mechanism of propulsion was clearly superior to the navy's paddlewheels, which were vulnerable to being disabled by enemy fire, but the admirals, and their advisers, were not impressed.

If the British were not interested the Americans were, and a second ship was built in 1838 with funding from a wealthy naval officer from New Jersey, named Robert F. Stockton. The 70 ft-long iron vessel, appropriately named *Stockton*, performed well, as reported in *The Times*, which said the tests were 'quite conclusive of the success of this important improvement in steam navigation . . .'. But not even *The Times* could persuade the ossific Admiralty of the superiority of screw propulsion, and Ericsson gave up the struggle. *Stockton* sailed to the United States in April, and

Ericsson followed seven months later aboard Brunel's *Great Western*.

Shortly after his arrival in New York Ericsson won a prize and gold medal for submitting the best fire-engine design. Regardless of this good start, the New World was not without its frustrations. Stockton, uncannily like Trevithick's Uville, turned out to be a much lesser man on his home turf than he had appeared in England. In 1841 Stockton commissioned Ericsson to design a screw-propelled iron warship for the US Navy. Completed three years later, the vessel, named *Princeton*, was remarkably successful and went on to have a long and distinguished record. But Stockton, who took all the credit, never even invited Ericsson aboard for the trials, and then held up the payments that were due to him from the navy. Quite aside from investing his time, Ericsson had spent $6,000 of his own money on the project. By the spring of 1846, with only $38.54 in the bank, he had still not received any payment for his work on *Princeton*.

Ericsson also rekindled his old flirtation with the flame, or caloric, engine, and in 1851 two investors paid $10,000 for a 10 per cent interest in his patent. That same year he persuaded some New York merchants to invest about $500,000 in building a ship powered by a caloric engine. The ship, named *Ericsson*, was 260 ft long with a displacement of almost 2,200 tons, and was launched in 1852. Her $130,000 caloric engine had four massive cylinders, each 14 ft in diameter. Aside from the dissenting voice of the *Scientific American*, which predicted its failure, the ship and its new engine were received with great excitement. There was every expectation that the caloric engine would one day supplant steam, but, as with *Novelty* more than two decades before, popular sentiment was to be disappointed and *Ericsson* fell short of the mark. The ship itself was stable and seaworthy but could not exceed a modest 8 mph and the high running temperature of the engine caused its rapid deterioration. Ericsson finally gave up on the caloric engine in 1855, still believing it would some day succeed.

Ericsson's most celebrated achievement, winning him the praise and adulation of his adopted country, came about through the Civil War. At the start of hostilities in 1861 Abraham Lincoln appointed a board to invite proposals for an ironclad warship. The Confederate forces were already going ahead with their plans for an ironclad frigate, the *Merrimac*, and the Unionists did not want to be at a military disadvantage. Ericsson responded by writing a stirringly patriotic letter to the President, sending him plans for an ironclad, 'for the destruction of the rebel fleet at Norfolk and for scouring the Southern rivers and inlets of all craft protected by rebel batteries'. His proposed ship was so simple to construct that it could be ready within ten weeks. He proposed it should take up a position 'under the rebel guns at Norfolk . . .', predicting 'that within a few hours the stolen ships would be sunk and the harbor purged of traitors'. Invited to Washington to meet the board, it took Ericsson two hours to make his case, and they took two hours to recommend acceptance. Within two months of writing to Lincoln the keel of the new ship, *Monitor*, was laid, and construction was finished within three months. During the trials there was such a problem with her steering that one newspaper ran a scathing article under the heading 'Ericsson's Folly'. All that was needed was an adjustment to the rudder's balance, but the naval authorities planned to dry-dock *Monitor* and fit a new one. This was all too much for a man who had suffered his fill of bureaucratic incompetence, and he flew into a rage. 'The *Monitor* is MINE,' he bellowed at the top of his stentorian voice, 'and I say it shall not be done.' The rudder was not replaced and the problem was quickly remedied.

Monitor was armed with two 11 in. guns mounted in a revolving turret, and clad in an impenetrable shield of iron, making her the most formidable warship afloat. But the Confederate's ironclad *Merrimac* was already on the loose and on 7 March, the day after *Monitor* left New York harbour, she sank two Union ships, the *Cumberland* and *Congress*, in Hampton Roads, off Norfolk, Virginia. A third ship, *Minnesota*, was badly damaged and sank the following day.

While *en route* to Hampton Roads *Monitor* ran into a heavy storm, but she handled remarkably well and her chief engineer, Alban C. Stimers, said the storm 'proved us to be the finest sea-boat I was ever in . . .'. Battle was joined with *Merrimac* on the morning of 9 March and continued for more than three hours. At one point *Merrimac* attempted to ram her assailant, but 'she is so flat and broad that she merely *slides* away from under our stern . . .'. If *Monitor's* commander had handled her better she would have sent *Merrimac* to the bottom in no time: 'give me that vessel', *Merrimac*'s commander, Catesby Jones, commented after the battle, 'and I would sink this one in twenty minutes.' Nevertheless, *Merrimac* broke off the engagement and withdrew, her command of the sea at an end.

Ericsson was inundated with congratulations and praise. 'Captain Ericsson, I congratulate you on your great success,' wrote the Chief Engineer on the day of the battle. 'Thousands here this day bless you. I have heard whole crews cheer you, every man feels that you have saved this place to the nation . . .'. The Senate and House of Representatives passed a joint resolution acknowledging

Ericsson 'for his enterprise, skill, energy and forecast . . . in the construction of his ironclad boat the *Monitor*' and thanked him 'for the great service which he has thus rendered to the country'. Congratulations flooded in from the rest of the Union and from around the world.

Military strategists recognised that Hampton Roads heralded a new era in naval warfare, and within one week of the battle Ericsson received a request for six more monitors. The first one had cost $275,000, generating a profit of $79,857.40, from which he received a quarter share. But the new monitors were to be bigger – 200 ft rather than 172 ft long – with construction costs of $400,000. Two additional monitors were ordered in the summer that were to be twice as long as *Monitor*, at a total cost of $1.5 million. Ericsson also received requests from Sweden, Norway and Chile. Fifty-seven-year-old Ericsson was set for the rest of his life. It was to be a long and busy life, and he kept working right up to within a few weeks of the end.

The beginning of the end came early in 1889, and was marked by a rapid deterioration in his health. He was eighty-five years old and his friends were so alarmed by the changes in his appearance that they wanted to call a doctor. Ericsson refused point-blank. The swelling in his limbs suggested dropsy – an accumulation of fluid in the tissues – and he instructed his faithful secretary to look up his symptoms in a medical treatise. The deterioration continued and when his fever and thirst increased, as his appetite declined, he agreed to see a doctor. Ericsson was suffering from an inflammation of the glomeruli, the minute clusters of capillaries through which the kidney filters blood. The condition, named Bright's disease after the physician who first described it, was caused by a bacterial infection. It would have been eminently treatable with penicillin, but that was not discovered until almost forty years later. The disease was fatal.

On 5 March, the anniversary of the day *Monitor* sailed from New York harbour, Ericsson was able to descend the stairs and take some food, but he dozed throughout most of the day and into the night. Then the moaning began. With some difficulty his secretary managed to persuade him to go to bed, and the following morning he sent for Ericsson's old friend and physician, Dr Thomas Markoe.

'How do you feel, Captain?' asked his old friend.

A smile of recognition flickered across the other man's face.

'Markoe,' Ericsson began after a pause, 'can a man who has Bright's disease do any more work?'

'Captain, a man who has Bright's disease has no right to do any work.'

Ericsson had always said that when his 'usefulness to mankind' was at an end he no longer wanted to live. He died two days later, on the anniversary of *Monitor*'s battle with *Merrimac*.

Ericsson's old business partner, John Braithwaite, was six years his senior and had a shorter and far less colourful life, predeceasing Ericsson by almost two decades. Like Robert Stephenson, his interests changed from building locomotives to building railways, and five years after Rainhill he gave up managing the engineering works to begin a new life as a civil engineer. His old partnership still continued for another three years, but failed in 1837. Fortunately for Braithwaite the bailiffs focused their attention on his junior partner and Ericsson had to spend some time in the Fleet prison as a foreign debtor.

Working with his old friend Vignoles, Braithwaite

surveyed the Eastern Counties Railway, and was appointed engineer-in-chief two years later, in 1834. He recommended laying the rails with a 5½ ft gauge track, later reduced to 5 ft. This was unsound advice, because eight years later the board, acting on the recommendations of Robert Stephenson, had the track replaced by standard gauge, at considerable expense and inconvenience. Braithwaite left the company and spent two years in France surveying railways. In addition to working as an engineer and surveyor, he co-founded the *Railway Times*, a popular publication dedicated to the booming new industry. After surveying Portsmouth's Langston Harbour in 1850, and building a brewery in Brentford on the outskirts of London the following year, he spent most of the remainder of his time as a consultant engineer. He died suddenly on 25 September 1870, aged seventy-three.

Soon after returning to work after the Rainhill trials, Timothy Hackworth was given a new project by the Stockton & Darlington Railway. The line was about to be extended to Middlesbrough and they wanted to build a new locomotive for passenger service. This was a radical departure for the company, whose engines had hitherto been built for hauling goods. What was needed was a light, fast locomotive, the opposite of what Hackworth was accustomed to, and he responded with a radical new design. The boiler had a straight-through flue, fitted with a series of small tubes, set at right angles, through which water circulated. The boiler barrel was no longer a prominent part of the engine, and since it was surrounded by a low wooden barrier to give safe access for the crew, it was essentially out of sight. Other innovative features included a spherical steam reservoir atop the boiler to prevent priming, a cranked axle and cylinders mounted inside the engine's frame. The new locomotive, appropriately named *Globe* for its large copper orb, was built by Robert Stephenson & Co.

Hackworth's grandson, Robert Young, claimed that the Stephensons' company had used *Globe* as a model for *Planet*, delaying its construction until their own new locomotive had been completed. While there may be some truth in the allegation, the only features the two engines shared were the cranked axle, which was nothing new, and internal cylinders. *Globe* appears to have operated successfully for nine years, until it exploded when the boiler was allowed to run dry. Regardless of *Globe*'s seemingly satisfactory service, no others of this design were built in its likeness, and there were no evolutionary descendants.

The fortunes of the Stockton & Darlington Railway improved considerably after the first few difficult years and profits steadily rose. Most of the business was still coal haulage, and Hackworth designed several new locomotives for this task. As they did not need to go any faster than about 6 mph, multi-tubular boilers, which were more expensive to build and maintain, were not needed. Instead, Hackworth used straight-through and return-flue boilers, often with some additional tubes to increase the heating surface. Most of these locomotives were large and heavy and were built on six wheels to reduce the stresses on the track.

Hackworth also undertook engineering jobs in his own time, an arrangement which was mutually agreeable to the company. Hackworth, now in his mid-forties, was becoming increasingly entrepreneurial. In 1833 he entered into a new relationship with the company: instead of being an employee he became a contractor, paying the company an annual fee for the use of their equipment, and receiving payment for the tonnage hauled. The financial arrangement was probably more in the company's favour than Hackworth's, and he did not make his fortune, but the new arrangement allowed him to pursue his business interests.

Entering into partnership with a man named Downing, they purchased some land in Shildon, where they erected a workshop and foundry for building locomotives and stationary engines. The new company of Hackworth & Downing, usually referred to as the 'Soho Works', was managed by Hackworth's brother, Thomas. They sold engines at home and abroad, and the Stockton & Darlington Railway was one of their biggest customers. Many of the locomotives Hackworth designed were built in his own factory, but many more were contracted out. Some had return-flue boilers, like *Samson*, built for the Canadian colliery. This massive and antiquated engine was typical of Hackworth's design, but he did build some multi-tubular locomotives, including one that was ordered by Russia. Completed in 1836, this locomotive was uncharacteristically light and racy. His son John, a capable lad of seventeen, was entrusted with travelling to Russia with the engine, accompanied by a small crew from the factory. Arriving late in the year, they had to travel overland by sleigh, encountering wolves along the way.

Hackworth finally severed his ties with the Stockton & Darlington Railway in 1840, having worked there for fifteen years. He was fifty-four years old and wanted to devote all his energies to his own business. For some years he had been thinking about building a locomotive that encompassed everything he had learned during a lifetime's experience, but the pressures of work prevented him from doing so. In 1848, having just completed a large contract for the London & Brighton Railway, he decided it was time. The construction was a labour of love, and he named his new locomotive *Sans Pareil No. 2*. This was an eminently suitable appellation for a locomotive that was built just two decades after Rainhill, and which represented the pinnacle of his locomotive achievements. The multi-tubular boiler, which was twice the length of its predecessor's, had 221 tubes, giving a heating surface of 1,105 sq. ft, some fifteen times greater than *Sans Pareil*'s. Building the boiler must have brought back bitter memories of rivets and leaking seams, but there was no chance of history repeating itself because the technical advances of two decades enabled him to weld the boiler seams. The firebox added another 83 sq. ft to the heating surface, for a total of 1,188 sq. ft. The 15 in. diameter cylinders were mounted inside the frame, transmitting power to the driving wheels by a cranked axle. She was a large engine, weighing almost 24 tons compared with just under five tons for the original, and was mounted on six wheels.

The most appropriate comparison for *Sans Pareil No. 2* would be one of Stephenson's long-boiler locomotives. The 'A' engine which took part in the gauge trials had the same weight, cylinder diameter and boiler length, though its diameter was 7 in. less. As the boiler was narrower there were fewer tubes, and the firebox was also smaller, giving *Sans Pareil No. 2* a total heating surface that was 26 per cent larger than that of the 'A' engine.

It would be remarkable if Hackworth had not entertained some 'what if?' thoughts as he and his son worked away on the magnificent new engine. What if he had been able to build *Sans Pareil* in his own workshop? Supposing he could have devoted his entire energies to the project? What if he had been able to have access to all the benefits and resources available to him now? And what if he could have charged past Mr Rastrick driving *this* engine? He probably shared some of these thoughts with his son, chatting away about the past as fathers and sons are wont to do. And on 25 October 1849, twenty years and twenty days after the eve of the trials, John Hackworth junior took up his pen and wrote to Robert Stephenson:

> It is now about 20 years since the competition . . . was played off at Rainhill . . . Your father and mine were the principal competitors. Since that period you have generally been looked to by the public standing first in the construction of locomotive engines. Understanding that you have now . . . a locomotive engine which is said to be the best production that ever issued from Forth Street works, I come forward to tell you publicly that I am prepared to contest with you, and prove to whom the superiority in the construction and manufacture of locomotive engines now belongs.

After discussing the enormous interest railway companies would have in such an experiment, he concluded:

> Relying upon your honour as a gentleman, I hold this open for a fortnight after the date of publication.
>
> I am, Sir, yours respectfully,
>
> John W. Hackworth.

The letter was published in several journals, including *Herapath's Railway Journal.*

Just imagine. A second chance to take on the might of the Stephensons. What a grand finale that would be to Timothy Hackworth's long and patient climb to the top. But it was not to be so, and the challenge was never accepted. Nonetheless, *Sans Pareil No. 2* was Hackworth's grand finale, because he died the following year, aged 63, after a short illness.

The most likely reason why Robert Stephenson did not accept John Hackworth's offer was that he was deeply engrossed in a much greater challenge of his own. 'I stood on the verge of a responsibility,' he wrote, after the event, 'from which, I confess, I had nearly shrunk.' His challenge was to build a bridge to carry the Chester & Holyhead Railway across the Menai Strait, which separates the Isle of Anglesey from the north coast of Wales. Bridging the quarter-mile gap so as not to interfere with shipping was a major problem. Telford had resolved this problem more than twenty years earlier by using a suspension bridge to carry a road across the strait. However, according to Smiles, Stephenson rejected doing the same because he thought such a bridge would be insufficiently strong and rigid. Stephenson wrestled with the problem for a considerable time before arriving at a likely solution. This was to build a box girder – essentially a wrought-iron tube – through which the railway would run. Such a bridge had never been built before and his reason for choosing this particular construction was that tubes are both strong and stiff. As there was no precedent for guidance he enlisted the help of one of his father's old associates, William Fairbairn, a practical engineer. Fairbairn built a series of tubular models with different cross-sectional shapes, including round ones, and tested them by supporting them at either end and loading them in the middle with weights until they failed. Among the conclusions from the experiments were that rectangular tubes were stronger than round ones; the height should be greater than the width; failure began at the top of the tube, which therefore had to be stronger than the bottom. Fairbairn also found that a tubular bridge would be sufficiently stiff and strong without the need of additional support from suspension cables.

Robert Stephenson was a cautious engineer but exercised special care in this project because of an incident that had occurred two years earlier, while working on the same railway. It was an engineer's worst nightmare: a structural failure causing loss of life. The disaster set him back hard on his heels, all but shattering his brittle self-confidence.

The Chester & Holyhead Railway, authorised in 1844, was a bold venture, which would ultimately link London with Holyhead, and its ferry to Ireland. Robert Stephenson was appointed engineer, and the eighty-five-mile route he

Construction work on one of the box girders of the Britannia bridge, which would span the Menai Strait.

chose presented him with some of the largest challenges of his career. First among them was crossing the River Dee, outside Chester. He initially planned building an arched bridge of bricks and masonry, but eventually decided to use iron. Cast iron was considerably cheaper than wrought iron and had been used successfully for the world's first iron bridge, at Coalbrookdale, so this is what he chose. In keeping with the bridges then being constructed on several railways, he decided to use girders, which are just a type of beam.

The River Dee was too wide to be bridged by a single span, so Stephenson designed for three, each of 109 ft. The middle span was supported at either end by a pair of vertical piers, standing in the river. These piers also bore one end of each of the shoreside spans, their inner ends being supported by abutments, built on the river banks. The girders were strengthened by wrought-iron suspension rods, called truss rods, positioned on either side of each girder. Such trussed girders had previously been used on other bridges, but only for shorter spans. The bridge was completed towards the end of September 1846, and inspected and certified by the Board of Trade the following month.

On the evening of 24 May 1847 the 6.20 passenger train out of Chester was crossing the bridge, hauling five carriages. The locomotive had cleared the first two spans and just reached the third when the driver heard an odd noise and noticed the bridge was vibrating far more than usual. He immediately opened the regulator fully, hoping to reach safety, but the locomotive barely reached firm ground when there was a loud crack as one of the girders failed. The five carriages plunged into the river, taking the tender with them and parting its coupling with the locomotive. The driver had a lucky escape, but his fireman, who was standing in the tender at the time, fell to his death. Four other people were killed, but most of the twenty-five passengers survived, though there were many injuries.

Robert Stephenson was devastated by the disaster, and appeared ready to accept responsibility, but the company's solicitor was adamant there would be no admission of liability as it would expose them to considerable financial damages. Stephenson had much more at stake than money and his reputation because if he were found criminally negligent he could be imprisoned. At the inquest the defence argued that the girder could not have failed unless it had been acted upon by some large and unexpected force. This force was attributed to the train having derailed, slamming it into the girder.

Charles Vignoles and Joseph Locke were among several expert witnesses who testified on Stephenson's behalf. That Vignoles should do this, given his long history of enmity towards George Stephenson, not only speaks well of the man but also for the high regard in which Robert was held. Robert also reconciled his differences with Locke, and they once again became the firm friends they used to be. The jurors were not persuaded by the derailment argument. But they were swayed by the expert witnesses, and by the fact that other trussed-girder bridges had been built that had not failed. Their verdict of accidental death vindicated Stephenson, at least as far as the law was concerned.

When the broken girder was subsequently recovered from the river, a break was discovered that was attributed to the way the truss rods had been attached. This was clearly a design error for which Stephenson was responsible. But his major mistake was in using cast iron in the first place, because it is so much weaker in tension and less predictable than wrought iron. Brunel and Locke always avoided using cast iron in railway bridges. Stephenson had learned a hard lesson and would never make the same mistakes again.

Brunel cancelled all his appointments to be with his friend at the Menai Strait. The two men stood side by side smoking cigars, looking down at the sea, trying to judge its mood. They had been there since early morning, along with a huge and expectant crowd, but the sea had been too rough to allow them to proceed. The plan seemed straightforward enough. One of the four box girders lay on floating pontoons, moored against the shore. All that Stephenson and the army of men under his command had to do was to set the girder adrift, manoeuvre it into position with ropes and capstans until it came to rest at the base of the two towers upon which it had to be raised, and secure it there. The wrought iron girder weighed 5,000 tons and was 460 ft long, which would make it exceedingly difficult to control in such a strong current. And that was only the first step, because it then had to be hoisted 200 ft above the water, though that would not be done for several months. This would have to be repeated three more times for the other girders.

Each girder was for one line of track, and the track was double, so the two main spans of the bridge were formed of paired girders. The central pier, in the middle of the strait, was built upon a rocky outcrop, Britannia rock, from which the bridge took its name. Long before construction had started on the girders a 75 ft-long model had been built and tested with increasing loads until it failed. A series of design improvements were then made until it could withstand twice the original load before failing – Stephenson was not going to make any mistakes this time.

He had tried floating the girder into position the previous day, 19 June 1849, but abandoned the attempt when one of the capstans failed, too late in the day to be repaired. Evening was beginning to fall again, and if this were not to be another disappointing day he would have to act very soon. There was still a stiff breeze blowing from the south-west, and a strong tide was running, but he decided to try.

Once he ordered the girder to be released from its mooring it carried smoothly with the tide, guided by a number of cables paid out from different vantage points along the shore. The spectators were delighted. But once it got further out into the channel its speed picked up alarmingly. Stephenson ordered it to be checked, but one of the thick cables parted, and it just kept on going.

The most critical phase of the entire operation was guiding the end of the girder so that it butted against the base of Anglesey tower. Once that had been safely secured it would be relatively straightforward to manoeuvre the other end towards the Britannia tower. The cable guiding the Anglesey end was controlled by a capstan on the Anglesey shore, and when the foreman saw the girder was going to miss the tower he ordered his men to bend to the task. As they worked at the capstan it jammed and the runaway girder gave the cable such a violent tug that the capstan was torn from its mounting, throwing men into the sea. Fortunately there was a second cable – Stephenson had doubled up on them all – and, seizing this, the foreman dragged it along the shore, shouting for the crowd's assistance. There was an immediate response as people rushed forward to take hold of the thick rope. They dug in their heels and bent their backs but the wayward girder dragged them along the shore. As more joined in, they gradually began to win the battle. At length the Anglesey end bumped against the tower and was lashed in place. The capstan controlling the other end was then set to work, and within a short time it was secure against the opposite tower. The crowds cheered, a cannon was fired and a band began to play. Stephenson could now retire to bed – he had not had a proper night's sleep for three weeks. 'Often at night I would lie tossing about,' he wrote, 'seeking sleep in vain. The tubes filled my head. I went to bed with them and got up with them.'

Hydraulic jacks were installed at the top of the towers for raising the girders by chains. At Stephenson's insistence each girder was raised only a few inches at a time, and a temporary column was continuously built up beneath it so that, in the event of an accident, it would fall only a few inches. His extreme caution was rewarded when one of the jacks failed, releasing the girder and sending fifty tons of chains and tackle crashing earthward, killing one of the workmen. Although the girder fell only nine inches the damage took six weeks to repair. The second girder was not raised until the beginning of 1850, and the bridge was not completed until towards the end of that year. The significance of this railway was that it linked London, the seat of government, with Ireland, via the ferry to Dublin.

Stephenson was now only forty-seven, and had more than enough money, but what he did not have was good health, or true happiness. He no longer had a wife to share his troubles: Fanny had died in the autumn of 1842, having been diagnosed with an incurable cancer two years earlier. Childless and very much in love, she had implored him to remarry when the time came, so he could have the son he always wanted. But he never remarried. Unable to go on living in a house with so many memories, he moved to a new home in Westminster and immersed himself in work. Although he put on a brave face for his friends, he was a lonely man, and with no family of his own he felt he had little to live for. He took great comfort from his friends and spent many a night dining out in their company, often staying up until quite late to enjoy one more cigar, one more conversation. Like his father he declined a knighthood, but he did become a Member of Parliament, representing the town of Whitby, on the Yorkshire coast, from 1847 until the day he died.

Stephenson was much in demand for his advice and was too good-natured to turn people away. As a consequence, he was always being hounded by railway entrepreneurs and speculators. They would often call at his office unannounced, and some thought nothing of waylaying him at his home. But when it all got too much for him he would escape aboard his yacht, *Titania*. She was a comfortable vessel, always well supplied with the best cigars and wines, and he was never happier than when setting off on one of his trips, usually accompanied by some of his friends. He always had time for his friends, no matter what.

One evening his biographer, Samuel Smiles, was visiting him at his London home when a note from Brunel arrived. Brunel's *Great Eastern* was nearing completion at Russell's shipyard on the Thames, and was about to be launched. There were serious difficulties because of her large size and Brunel asked if Stephenson could come down to Blackwall early the next morning to give him the benefit of his advice.

Stephenson arrived just after six o'clock on a cold, dark morning. He had not bothered to wear his greatcoat, which was not unusual, and wore only thin boots. His moral support was as valuable to his old friend as his advice, and they spent the whole day together, smoking cigars and discussing the launch. At around noon the baulk of timber upon which Stephenson was standing tipped up, pitching him into the soft Thames mud up to his middle. Everyone urged him to get some dry clothes 'but, with his usual disregard of health, he replied, "Oh, never mind me; I'm quite used to this sort of thing;" and he went paddling about in the mud, smoking his cigar, until almost dark . . .'.

In the autumn of 1857 Stephenson took a sentimental journey to Northumberland to visit places once so familiar to him when he was growing up. He was not quite fifty-four but his health was failing and when he visited Killingworth he must have known it was for the last time. Arriving at the small cottage where he had spent most of his boyhood he stood outside, gazing up at the sundial over the door that he had helped his father build. The current tenants welcomed him inside, and he was touched to see that the combined bookcase and writing desk his father had made was still there. He asked the woman if she knew there was a secret drawer in the desk, but she was adamant it had no such thing. 'Oh, but it has,' Robert replied, stepping towards the desk. 'I know it has: for it was made by my father.' Pressing the unseen button, the secret drawer flew open, but, to everyone's disappointment, it was empty. They stayed there talking for a few minutes more and when it was time to leave there were tears in his eyes.

He walked down the road to the local alehouse, the Closing Hill House, to see if his old friend Robert Tate still lived there. Stephenson was pleased to find he did, but there was no sign of recognition when he stepped into the parlour.

'What, don't you know me, old friend?' asked Robert, much affected.

'Why—' said Tate, after a pause, 'it must be Robert Stephenson.'

'Ay, my lad,' answered Robert; 'it's all that's left of him.'

That same autumn he set off aboard *Titania* for a trip to

Scotland, accompanied by George Bidder, whom he had met at Edinburgh University, and two other friends. One of them recalled life aboard *Titania*. She had:

> a crew of sixteen men, a good cook, and a first-rate cellar . . . We had an ample supply of astronomic and mathematical instruments; and one person on board, at least, knew how to use them . . . There was a capital library . . . and . . . each man before going to bed selected a book. At seven in the morning a cup of coffee was served in bed . . . [we] then read or snoozed till nine . . . [Their itinerary included Loch Ness, where they made] repeated observations on the temperature of the water at the surface, and at various depths, at one place . . . at a depth of 170 fathoms [1,020 ft].

On the return journey down the west coast they visited the Britannia Bridge, and Stephenson gave them a guided tour:

> I can never forget the interest which [he] excited in us, as he described in his quiet way the general design . . . the difficulties encountered and overcome in the erection . . . The principal part of the description was given on the top of the tube, on a beautiful morning . . . We smoked a cigar in quiet contemplation before we left the spot, none of the party being disposed to speak . . .

The following autumn *Titania* headed for warmer waters, with Stephenson and another small party of friends aboard. Stopping off at Algiers and Malta along the way, they arrived in Alexandria on 3 November 1858, dropping anchor opposite the Pasha's palace. Stephenson was extremely well read on the subject of Egypt, and delighted to share his knowledge with the rest of the party. 'The climate,' Stephenson wrote, 'is perfectly delicious . . .' and he was feeling 'in good spirits . . . and actually recovering my flesh.' Before leaving for Cairo, where they would be spending Christmas, he received a letter from Brunel. His old friend and his wife were visiting Cairo for the holidays and invited him to dine with them on Christmas Day. Brunel, like Stephenson, was also a sick man. He was suffering from Bright's disease, and easily fatigued. The Brunels were staying at Cairo's Hôtel d'Orient, and he could be seen riding around the streets of the city on a donkey, 'as free from care as a schoolboy'.

As planned, the two old friends, the greatest engineers of the era, dined together on Christmas Day, each delighting in the company of the other. And as they sat there chatting into the night, enjoying just one more cigar together, they must have known there was little time left for either of them.

Stephenson spent a few weeks in Paris on his way home, arriving back in London early in February. The trip abroad had lifted his spirits and he enjoyed 'an unaccustomed sense of vigour' but the 'deep-seated mischief in liver, stomach, and nerves' remained unchanged. By late summer his health had deteriorated, and before embarking on *Titania* to sail for Norway, he revised his will. The trip was made so he could attend a special dinner in his honour, and as his party was larger than usual, his friend Bidder took some of the guests aboard his own yacht, *Mayfly*. Thomas Gooch was one of the guests aboard *Titania*, along with George Phipps, who had been Stephenson's draughtsman at the Forth Street Works many years before.

After the dinner Stephenson was warmly praised for bringing steam locomotion to Norway. Some time before he had been decorated with the Grand Cross of the Order of St Olaf, an honour normally restricted to Norwegian citizens. He had intended responding with a discourse on the rise and progress of the railway, but when he went to stand he was overcome by an attack of nausea and faintness. Unable to deliver his original address, he replied with an eloquent speech in which he deflected to others the praise that had been heaped upon himself. His self-effacing offering, so typical of the man, was to be his last public speech.

By the following evening he was so ill that his friends thought he might die before reaching home, so they set sail early the next morning. Running into bad weather the two yachts became separated, and spent a week battling heavy seas, not meeting up until they made a landfall off the Suffolk coast, on 13 September. The sight of England seemed to rally Stephenson, and he insisted on walking to the nearby railway station rather than being carried.

Once back in his own home his condition improved, and on one occasion he surprised everyone by appearing in his drawing room, announcing he had grown tired of being confined to bed. But it was a short-lived recovery, and the following day he was gravely ill, and the 'obstinate congestion of the liver was followed by dropsy of the whole system . . .'. He died shortly before noon on 12 October 1859, one month after his friend Brunel.

News of his death seemed to sadden the entire country. 'Never before,' wrote his biographer, 'had the death of a private person struck so deeply the feelings of his country.' He was buried in Westminster Abbey, alongside other heroes of the British Empire.

Three years before his death he delivered his inaugural address as President of the Institution of Civil Engineers. Choosing the subject of the impact of railways upon the country, he shared some revealing figures with his audience. There were over 8,000 miles of rail, serving 2,416 stations, and carrying 300,000 passengers every day and over 111 million each year. The annual revenues generated exceeded £20 million (£1,140 million) and there had been 'no instance, in the annals of any railway, where the annual traffic has not been of continuous growth'. Capital expenditure had amounted to at least £286 million (£16,302 million), which was more than 'four times the amount of the annual value of all real property' of the country, and 'more than one-third of the entire . . . National Debt'. The railway companies employed no fewer than 90,409 workers, and annually burned 1.3 million tons of coke, and used 20,000 tons of iron and 26 million sleepers to maintain the track. The wood for the sleepers was equivalent to the timber from a 5,000-acre forest.

In concluding his address he said it was:

> my great pride to remember, that whatever may have been done, and however extensive may have been my own connection with railway development, all I know and all I have done, is primarily due to the Parent whose memory I cherish and revere.

Notes

1 THE SPORT OF KINGS

p.2 The reference to a dalmatian dog, attributed to John Hackworth in 1875, is given in R. Young, *Timothy Hackworth and the Locomotive* (Shildon: 'Stockton & Darlington Railway' Jubilee Committee, 1975; first published 1923), pp. 138–9.

p.4 The prices of coal are from ibid., p. 121.

p.4 and throughout. The current values of money are based upon data from R. Twigger, *Inflation: The Value of the Pound, 1750–1998* (Research paper 99/20, House of Commons Library, 1999).

p.4 The reference to share prices is from Young, *Timothy Hackworth*, pp. 118–20.

p.5 Walker's report on the respective merits of fixed and locomotive engines was published as J. Walker, *Report to the Directors on the Comparative Merits of Locomotive and Fixed Engines as a Moving Power* (London: Liverpool & Manchester Railway, 1829), PRO, ZLIB 4/57.

p.5 In his running-cost estimates Walker calculated that fixed engines would cost 0.0653*d* less per mile per ton of goods than would locomotives.

p.6 A transcript of Stephenson's critique of the consultants' report is available at the Public Record Office (PRO), RAIL 1008/88/1.

p.6 The 13 April 1829 meeting of the board is recorded in the minutes, PRO, RAIL 371/1, p. 112.

p.7 The board's decision to offer a £500 premium was made during the 20 April 1829 meeting (PRO, RAIL 371/1, p. 113).

p.7 There has been some discrepancy in the year of Booth's birth. See H. Booth, *Henry Booth, Inventor – Partner in the* Rocket *and the Father of Railway Management* (Ilfracombe: Stockwell, 1980), p. 13.

p.8 Booth's quote about the fanciful ideas submitted is from S. Smiles, *Memoir of the late Henry Booth* (London: Wyman, 1869), pp. 37–8.

p.9 *The Leeds Mercury*, 10 October 1829.

p.9 Very little is known about Burstall. The best, albeit short, account is given in C. F. Dendy Marshall, *A History of the Railway Locomotive down to the End of the Year 1831* (London: Locomotive Publishing Co., 1953), pp. 186–92.

p.10 The report of the accident to Burstall's locomotive is contained in the minutes of the 5 October 1829 meeting of the board of the Liverpool & Manchester Railway Company, PRO, RAIL 371/1, p. 137.

p.12 The report on 'The great lightness . . .' is from *Mechanics' Magazine* 322 (10 October 1829), p. 115.

p.13 Hackworth's early experience with locomotives is discussed in Young, *Timothy Hackworth*, pp. 47–52.

p.15 Dionysius Lardner's public lecture, when he referred to George Stephenson as 'the father of the locomotive engine', was referred to in a letter of rebuttal, written to Lardner by William Hedley (another contender for the title) on 10 December 1836. This letter is quoted in O. D. Hedley, *Who Invented the Locomotive Engine?* (London: Ward Lock, 1858), pp. 34–7.

p.15 The quote about inventors is from S. Smiles, *The Life of George Stephenson, Railway Engineer* (London: John Murray, 1857), p. 501.

p.17 The comment about Robert Stephenson's charm is attributed to F. R. Conder and quoted in L. T. C. Rolt, *George and Robert Stephenson: the Railway Revolution* (London: Longman, 1960; reprinted London: Penguin Books, 1984), p. 230.

p.19 The requirement that the locomotive should 'consume its own smoke' is one of the stipulations of the Act of Parliament, *An Act for making and maintaining a Railway or Tramroad from the Town of Liverpool to the Town of Manchester, with certain Branches therefrom, all in the County of Lancaster, 5th May 1826*, Science Museum Library, MSS 605.

p.21 The report of the 'simultaneous burst of applause' which followed the appearance of the different carriages in the forenoon was given in *Gore's General Advertiser*, cited in Booth, *Henry Booth*, p. 73.

p.21 The report about 'shooting past the spectators . . .' is from *The Liverpool Courier*, 7 October 1829. There is some discrepancy between contemporary accounts as to whether *Rocket* was first demonstrated with or without a load. Both the judges' report, and that given by Stephenson and Lock, state that she first ran hauling a load: J. U. Rastrick, N. Woods and J. Kennedy, 'Report of the Competition on the Liverpool & Manchester Railway, to the Directors of the Liverpool & Manchester Railway' (1829), transcript from the Science and Society Picture Library, NRM/RH1/B460201B–B460210B; R. Stephenson and J. Locke, *An Account of the Competition of Locomotive Engines at Rainhill, in October, 1829, and of the Subsequent Experiments*, Liverpool: Liverpool & Manchester Railway Company, PRO, ZLIB 4/57.

p.21 During her trials before Rainhill, the *Rocket* did not exceed a speed of 12 mph (letter from R. Stephenson to Henry Booth, 5 September 1829, part of PRO, RAIL 1088/88). It appears that none of the other competitors had the opportunity of testing their locomotive before their arrival at Rainhill. While they may have taken some short test runs at Rainhill before 6 October it is exceedingly unlikely they reached any good speeds.

p.22 Stannard's boyhood recollection of his ride aboard *Rocket* was given in a letter he wrote to *The Engineer* 58 (17 October 1884), pp. 302–3.

p.22 The quote 'The faults most perceptible . . .' is from *Mechanics' Magazine* 322 (10 October 1829), p. 115.

p.22 The initial report of *Rocket*'s smoke emission was clarified in the *Mechanics' Magazine* 324 (24 October 1829), p. 152.

p.23 The quote about the landlady of the Railroad tavern is from *The Times*, 8 October 1829, p. 3.

p.25 The quote 'Almost at once . . .' is from *The Mechanics' Magazine* 322 (10 October 1829), p. 116.

p.25 The quote 'It seemed indeed to fly . . .' is from *The Leeds Mercury*, 10 October 1829, p. 3.

p.25 *The Liverpool Chronicle*, 10 October 1829, p. 327.

p.27 B. Cleverton, 'New gas power-engine', *Mechanics' Magazine* 137 (1826), pp. 385–9.

p.27 The quote 'Mr Brandreth's . . .' is from *The Liverpool Courier*, 7 October 1829, p. 320.

p.27 *Mechanics' Magazine* 322 (10 October 1829), p. 115.

p.29 Data for the Stockton & Darlington railway about horses and locomotives hauling coal wagons are from C. E. Lee and K. R. Gilbert (eds), *Railways in England, 1826 and 1827* (Cambridge: Newcomen Society, 1971), a translation of C. von Oeynhausen and H. von Dechen, *Ueber Schienenwege in England. Bemerkungen gesammelt auf einer Reise in den Jahren 1826 und 1827* (1829).

p.29 During the year 1828, horses hauled a total of 55,235 tons (43 per cent), compared with 74,550 tons for locomotives. Data from W. W. Tomlinson, *The North Eastern Railway: its Rise and Development* (London: Longman, 1914).

p.29 The quote 'no great velocity . . .' is from *The Liverpool Courier*, 7 October 1829, p. 320.

p.30 The quote 'after the most . . .' is from Rastrick *et al.*, '*Report of the Competition*'.

p.31 A copy of the company's stipulations for the contestants, as drawn up by the directors and dated 25 April 1829, was made available to the three judges on Monday 5 October 1829. This formed the preface to their report of the competition, and the following is a transcription of these original stipulations:

Stipulations & Conditions on which the Directors of the Liverpool & Manchester Railway offer a Premium of £500 for the most improved Locomotive Engine

1 The said engine must effectually consume its own smoke according to the provisions of the Railway Act of George IV.
2 The engine if it weighs six tons must be capable of drawing after it, day by day, on a well constructed railway on a level plane a Train of Carriages of the gross weight of twenty tons including a Tender and Water Tank at the rate of ten miles per hour with a pressure of steam in the boiler not exceeding 50 pounds per square inch.
3 There must be two safety valves, one of which must be completely out of the reach of contact of the engine man and neither of which must be fastened down while the engine is working.
4 The engine and boiler must be supported on springs and rest on six wheels and the height from the ground to the top of the chimney must not exceed 15 feet.
5 The weight of the machine with its compliment of water in the boiler must not exceed six tons and a machine of less weight will be preferred if it draw after it a proportionate weight and if the weight of the Engine & c [etc.] do not exceed five tons then the gross weight to be drawn need not exceed fifteen tons and that proportion for machines of still smaller weight provided that the engine can still be on six wheels unless the weight as above be reduced to four tons and a half or under in which case the Boiler & c may be placed on four wheels. And the Company shall be at liberty to put the boiler, Fire tube, Cylinders & c to the test of a pressure of water not exceeding 150 pounds per square inch without being answerable for any damage the Machine may accrue in consequence.
6 There must be a mercurial gauge affixed to the machine with Index not showing the steam pressure above 45 pounds per square inch and constructed to blow out at a Pressure of 60 pounds per square inch.
7 The Engine to be delivered complete for trial at the Liverpool end of the railway not later than the 1st October next [this was later extended to 6 October].

8 The price of the Engine which may be accepted not to exceed £550 delivered on the Railway and any engine not approved to be taken back by the owner. NB the Railway Company will provide the Engine with water and a supply of fuel for the experiment. The distance within the Rails is four feet eight inches and a half.

(Rastrick *et al.*, 'Report of the Competition'.)

2 LESSONS FROM THE PAST

p.32 The quote about 'only one man in three . . . ' is from 'Notes and incidents', dictated to Oxtoby by George Graham, who drove *Locomotion*, PRO, RAIL 667/427.

p.33 The quote about lubrication is also from 'Notes and incidents'.

p.33 The high incidence of wheel breakages on the Stockton & Darlington Railway is reported in R. Young, *Timothy Hackworth and the Locomotive* (Shildon: 'Stockton & Darlington Railway' Jubilee Committee, 1975; first published 1923).

p.34 An account of James Stephenson's unruly reputation is given in Young, *Timothy Hackworth*, pp. 299–300.

p.35 The account of the explosions of *Diligence* and *Locomotion* are based upon Rastrick's notebook, located at Senate House Library, London University, MS 155, and Hackworth's notebook, Science Museum Library, London, MS 598.

p.35 Hackworth's brief note on the death of Cree is from Hackworth's notebook, Science Museum Library, MS 598.

p.36 The quote about the mangled driver of *Salamanca* is from *The Leeds Mercury*, Saturday 7 March 1818.

p.38 By wrapping the end of a drinking straw in a narrow band of folded paper towel, it can be made to fit snugly into the spout of a kettle. The straw can be used to deliver a jet of steam into the narrowest of openings. Plastic bottles and cardboard milk cartons can then be used to demonstrate the effects of atmospheric pressure.

p.40 Newcomen's 1712 engine was known as the Dudley Castle engine, because it was erected within sight of that landmark. A working replica, completed in 1986, can be seen, sometimes in steam, at the Black Country Museum, Dudley. See J. S. Allen, 'The 'Dudley Castle', 1712, Newcomen engine replica, Black Country Museum, Dudley, West Midlands', *Transactions of the Newcomen Society* 69, 2 (1997), pp. 283–98.

p.40 The number of Newcomen engines is from L. T. C. Rolt and J. S. Allen, *The Steam Engine of Thomas Newcomen* (Ashbourne: Landmark Publishing, 1997), p. 124.

p.41 The first quote from Watt is taken from J. P. Muirhead, *The Life of James Watt, with Selections from his Correspondence* (New York: Appleton, 1859), p. 60.

p.41 The quote from Watt about his idea of a separate condenser is from ibid., p. 64.

p.43 Stories of Trevithick's remarkable strength are given in F. Trevithick, *Life of Richard Trevithick, with an Account of his Inventions* I (London: Spon, 1872), pp. 54–5.

p.45 It is unclear how secretive were the model steam engine experiments in Redruth. According to Trevithick, *Life of Richard Trevithick* I, Murdock conducted his experiments in secret. However, there is a letter written by Murdock's son John to James Watt, Jr (son of James Watt), in which it is stated that 'The model of the wheel carriage was made in 1792 [this should have read 1784] and was then shewn to many of the inhabitants of Redruth.' The letter is quoted in H. W. Dickinson and A. Titley, *Richard Trevithick: the Engineer and the Man* (Cambridge: Cambridge University Press, 1934), p. 46.

p.45 For a brief account of Murdock's accomplishments see the *Proceedings of the Institution of Mechanical Engineers* 2 (1850), pp. 16–25.

p.47 In the letter written by Murdock's son, already mentioned in note to p. 45, he states that Trevithick, accompanied by Vivian, called at his father's house about two years after the model was built (i.e. around 1786) to discuss the removal of an engine from a mine. During their visit they asked to see the model.

p.48 Francis Trevithick points out that the earliest of his father's account books in his possession was dated 1797 and that during the several years the book was in use not a single account was balanced. See Trevithick, *Life of Richard Trevithick* I (1872), p. 63.

p.48 Davies Gilbert was formerly known as Davies Giddy. In 1816 he inherited property from his father-in-law, at which point he adopted his wife's maiden name.

p.49 The quote about the few small hand lathes is from Trevithick Jr's experience of working at the foundry thirty years later, and is given in ibid. I, p. 107.

pp.49–50 The quote from Stephen Williams about the first trial of the Camborne carriage is from ibid. I, pp. 107–8.

p.50 The quote about the loss of the Camborne carriage is from ibid. I, p. 117.

p.51 The quote about the 1½ in. thickness of the boiler, and being able to sell his engines are from a letter from Trevithick to Gilbert, written 22 August 1802, quoted in ibid. I, pp. 153–5.

p.52 The account of being pelted by cab drivers is from ibid. I, p. 111.

p.53 The quote about the abandonment of the London steam carriage was made by Trevithick's partner Vivian, and is quoted in ibid. I, p. 140.

p.53 Trevithick's quote about his prospects is in a letter written to Gilbert from Bristol, 2 May 1803, cited in full in ibid. I, p. 158.

p.53 The quote about Boulton & Watt doing their utmost to report the explosion, and Trevithick's quote about safety valves are contained in a letter written to Gilbert from Pen-y-darren, 1 October 1803, cited in full in ibid. II, pp. 124–6.

p.54 The quote about Mr Homfray taking him by the hand is from the same letter, ibid.

p.54 Boulton & Watt's attempt to banish high-pressure engines is discussed in a letter written by Trevithick, quoted in part in ibid. II, p. 131.

p.54 A brief account of how the Pen-y-darren locomotive was built is given in an article in the *Mining Journal*, 2 October 1858, quoted in ibid. I, pp. 177–8.

p.54 The comment about the locomotive working very well is contained in a letter from Trevithick to Gilbert, written from Pen-y-darren, 15 February 1804, quoted in ibid. I, pp. 159–60.

p.56 The quote about steam discharged from the engine being turned up the chimney is contained in a letter from Trevithick to Gilbert, written from Pen-y-darren, 20 February 1804, quoted in ibid. I, pp. 160–1.

p.56 Trevithick's quote about the tone of the public being much altered is contained in the letter to Gilbert written from Pen-y-darren, 22 February 1804, quoted in ibid. I, pp. 161–2.

p.56 The prediction about the number of horses was contained in a letter received from a correspondent in Merthyr Tydfil, *The Cambrian*, 24 February 1804, reproduced in C. F. Dendy Marshall, *A History of the Railway Locomotive down to the Year 1831* (London: Locomotive Publishing Co., 1953).

p.56 The quote about the Pen-y-darren locomotive breaking the tracks was made by Rees Jones in the *Mining Journal*, 2 October 1858, given in Trevithick, *Life of Richard Trevithick* I, pp. 177–8.

p.57 An extract of Vivian's accounts to the end of November 1804 is given in ibid. I, pp. 228–9.

p.57 The quotation about running against horses for a wager appeared in *The Times*, 8 July 1808. There has been some discussion on the accuracy of the well known illustration depicting the circular track with the steam locomotive that Trevithick exhibited in London. See Loughnan St L. Pendred, 'The mystery of Trevithick's London locomotives', *Transactions of the Newcomen Society* 1 (1922), pp. 34–49. The same author also questioned the reliability of the illustration of the locomotive.

p.59 Trevithick's comments about the unrealised importance of the locomotive is contained in a letter he wrote from Camborne, 26 April 1812, to Sir John Sinclair. The entire letter is quoted in Trevithick, *Life of Richard Trevithick* II, pp. 41–3.

p.61 Teague's comment to Uville about his cousin is quoted in ibid. II, pp. 197–8.

3 THE LONDON CHALLENGE

p.67 New revelations on *Novelty*'s air pump were made when the replica was being renovated for a re-enactment of the Rainhill trials which took place in Wales in 2002. At the time of writing this new information had not yet been published.

p.67 *Albion*, 12 October 1829, p. 325.

p.68 The description of the workings of the *Novelty* appeared in two consecutive issues, alongside coverage of the trials in general: *Mechanics' Magazine* 323 (17 October 1829), pp. 129–42; 324 (24 October 1829), pp. 145–52.

p.68 Boiler dimensions for *Novelty* and *Sans Pareil* are from C. F. Dendy Marshall, 'The Rainhill locomotive trials of 1829', *Transactions of the Newcomen Society* 9 (1928–29), 1930.pp. 78–93.

p.71 Ericsson's flame engine was described in a manuscript entitled 'A Description of a new Method of employing the Combustion of Fuel as a Moving Power', which is in the archives of the Institution of Civil Engineers, archive No. 119.

p.71 For more information on Ericsson's heat engines see M. Lamm, 'The big engine that couldn't', *American Heritage of Invention and Technology* 8, 3 (1993), pp. 40–7.

p.72 Ericsson's comment on his fortune in meeting Braithwaite was contained in a letter he wrote to the *London Builder*, 23 April 1863, quoted in Church (1906), volume 1, p. 38.

p.72 A brief account of Braithwaite's life is given in Institution of Civil Engineers *Minutes of Proceedings* 31 (1870–71), pp. 207–11. An account of the loss of the *Earl of Abergavenny*, available at a Web site (http://www.weymouthdiving.co.uk/abergavenny.htm), places the value of the cargo at £74,000.

pp.74–75 An account of Braithwaite and Ericsson's fire engine is given in Church, *The Life of John Ericsson* (1906), pp. 44–8.

p.75 The quotation about *Novelty* drawing her load with ease is from *Mechanics' Magazine* 323 (17 October 1829), p. 135.

p.75 The attribution of comments about George Stephenson's view of the *Novelty* as a rival is given in O. J. Vignoles, *Life of Charles Blacker Vignoles, FRS, FRAS, MRIA, &c, Soldier and Civil Engineer* (London: Longman, 1889), p. 129.

p.76 That *Novelty* broke down during the second day of the trials owing to the breakage of her bellows is reported in two reliable sources: J. U. Rastrick, N. Wood and J. Kennedy, 'Report of the Competition on the Liverpool & Manchester Railway to the Directors of the Liverpool & Manchester Railway Company' (1829). The photograph of the original hand-written report and a transcript are in the Science and Society Picture Library, Science Museum, London, NRM/RH1/B460201B–B460210B; 'Observations on the Comparative Merits of Locomotive and Fixed Engines as applied to Railways, being a Reply to the Report of Mr James Walker, to the Directors of the Liverpool & Manchester Railway, compiled from the Reports of Mr George Stephenson, with an Account of the Competition of Locomotive Engines at Rainhill, in October 1829, and of the Subsequent Experiments by Robert Stephenson and Joseph Locke, Civil Engineers, Liverpool, PRO, ZLIB 4/57, pp. 60–83. This last source also records that the failure occurred during the second leg (p. 68).

p.76 Stephenson's comment about *Novelty* lacking guts is quoted in Vignoles, *Life of Charles Blacker Vignoles*, p. 130, and elsewhere.

p.77 The accident with the horse is reported in *The Times*, Monday 12 October 1829, p. 3.

p.78 The report on the smoke consumption of *Novelty* appeared in the *Mechanics' Magazine* 323 (17 October 1829), pp. 135–6.

p.78 The report in *The Albion* appeared on 12 October 1829, p. 325.

p.78 That Vignoles wrote Rainhill articles for the *Mechanics' Magazine* is confirmed by his son in his biography of his father (Vignoles, *Life of Charles Blacker Vignoles*, p. 128).

4 A MAN OF PRINCIPLES

p.80 The process of *corrosion* could be speeded up in wrought iron by daubing the leaks with corrosive materials like iron filings mixed with urine.

p.81 Arc welding may have been discovered by Sir Humphrey Davy as early as 1810, but was not available for industrial use until 1881. Gas welding seems to have begun in the early nineteenth century: in 1820 a German designed a pipe to burn a compressed mixture of oxygen and hydrogen – an extremely dangerous mixture. An oxygen/coal gas blowpipe was invented in 1838. (Science Museum Library, MS 1179, history of welding, dated 20 August 1975.)

pp.81–82 Cast iron and wrought iron are both derived from pig iron, the raw product of the blast furnace. Knowing what goes on inside a blast furnace is helpful in understanding the differences between these various iron products. Iron is the second most abundant metal in the earth's crust, and two of its commonest ores are different forms of iron oxide. It also occurs in combination with carbon and oxygen (as a carbonate), and with silicon (sand is an oxide of silicon). During the smelting process taking place inside the blast furnace, the iron ore is broken down, releasing it from the other elements. The blast furnace is a tall vessel with an opening at the top through which the ore is fed, mixed with coke, together with some limestone. Compressed air is blasted into the lower part of the chamber through a series of jets, and this causes the coke to burn incandescently hot. Coke is carbon, and forms carbon monoxide when it burns. This gas immediately combines with the oxygen in the iron ore to produce carbon dioxide. Freed from its combined oxygen, the molten iron sinks to the bottom of the furnace. The purpose of the limestone is to combine with the silicon in the ore to form a slag (calcium silicate), and this molten impurity also sinks to the bottom. As slag is less dense than iron, it floats on the surface of the molten metal, and is drawn off from time to time through a tap hole part-way up the side of the furnace.

The iron is periodically tapped off from the bottom of the furnace. The molten iron is channelled into a series of sand moulds, where it solidifies into ingots, called *pigs*. Pig iron contains about 5 per cent of impurities, mostly carbon, with small amounts of other elements, including silicon. Cast iron is made by remelting the pig iron, and therefore contains all the same impurities. The process commonly used to produce wrought iron at the time of the Rainhill trials was *puddling*, which took place in small hand-operated furnaces. A small quantity of pig iron was melted and stirred with an iron pole. During the process, which took nearly two hours to complete, most of the carbon in the iron combined with oxygen and was thereby removed. This reduced the carbon content from about 4 per cent to less than one-tenth of that amount. When the puddler judged the time was right, the ball of iron, which had the consistency of toffee, was removed from the furnace. It was then beaten with a hammer, or squeezed between rollers, to express the slag. The quality of the resulting wrought iron was usually enhanced by cutting it into slabs, piling these on top of one another, and heating the pile in a furnace to weld them together. The fused pile was then hammered or rolled into a slab. This process would be repeated several times, the effect being to increase the tensile strength of the wrought iron. It also caused the remaining inclusions to become elongated, giving wrought iron its characteristic fibrous structure. This fibrous structure gives wrought iron a finely striated appearance, something like wood grain, seen in well weathered wrought iron, like old ships' anchors and rusted fittings on old buildings.

p.82 The warning about boiler plate from Bedlington is given in a letter from George Stephenson to Robert Stephenson, 8 January 1828 (Science Museum Library, MS 1149).

p.82 The reference to the boiler plate used in *Rocket* is from R. Stephenson & Co. Ledger, 1821–31, R. Stephenson & Co. Collection, National Railway Museum, fo. 210, entry for 11 August 1829, cited in M. R. Bailey and J. P. Githero,

The Engineering and History of Rocket (York: National Railway Museum, 2000), p. 77.

p.82 Dixon's comment about Hackworth's ill humour at Rainhill is contained in a letter he wrote on 16 October 1829 (Science Museum Library, MS 1573).

p.83 The governing body of the Stockton & Darlington Railway Company referred to themselves as a committee rather than a board.

p.84 Hackworth's letter to Pease was written on 9 May 1825 and also dated 20 May 1825 (PRO, RAIL 667/1158/3).

pp.84–85 The letter of recommendation for Hackworth was written by Mr Potter, part owner of the Walbottle colliery, on 3 February 1825 (PRO, RAIL 667/1158/2). Although Potter said Hackworth had worked for him for eleven years, the earliest he started was late in 1815, so his tenure at Walbottle was closer to nine years.

p.85 Details of Hackworth's salary are contained in a note he wrote of expenses for the quarter ending 26 December 1827 (PRO, RAIL 667/1158/12). The rent and heating coal information is included in his notes covering the period 1825–49 (PRO, RAIL 667/1158).

p.85 Hackworth's father's comment about his son's aptitude is cited in R. Young, *Timothy Hackworth and the Locomotive* (Shildon: 'Stockton & Darlington Railway' Jubilee Committee, 1975; first published 1923), p. 43.

p.87 The quote about Blackett not being an engineer is from O. D. Hedley, *Who invented the Locomotive Engine?* (London: Ward Lock, 1858), p. 18.

p.88 Trevithick's patent is reproduced in F. Trevithick, *Life of Richard Trevithick, with an Account of his Inventions* I (London: Spon, 1872); the section referring to wheels appears on p. 133 of that volume.

p.88 That the Pen-y-darren locomotive had smooth wheels is contained in a letter written by Rees Jones, many years after he helped him build it. The letter, which was published in the *Mining Journal* on 2 October 1858, is quoted in its entirety in Trevithick, *Life of Richard Trevithick* I, pp. 177–8.

p.88 Although Young (*Timothy Hackworth*, p. 45) gives the year of the test carriage as 1811 there is some uncertainty, and it seems more likely to have been 1812. See A. Guy, 'North-eastern locomotive pioneers, 1805–27: a reassessment' in A. Guy and J. Rees (eds), *Early Railways: a Selection of Papers from the first International Early Railways Conference* (London: Newcomen Society, 2001).

p.88 Blenkinsop's rack-and-pinion system was primarily designed to reduce the weight of the locomotives without inducing slippage, rather than to overcome the perceived traction problem.

p.90 The report in the *Leeds Mercury* of 27 June 1812 is quoted in Young, *Timothy Hackworth*, pp. 61–2.

p.90 The expected saving of Blenkinsop's locomotive was reported in a letter to the *Monthly Magazine* for June 1814, written by Blenkinsop and quoted in C. F. Dendy Marshall, *History of the Railway Locomotive down to the End of the Year 1831* (London: Locomotive Publishing Co., 1953), p. 37.

p.90 According to Dendy Marshall (ibid., p. 81) an advert by 'Thom. Waters & Co., manufacturers of Trevithick engines', appeared in the *Newcastle Courant* on 3 June 1815.

p.91 The year 1813 seems the most likely for the building of the Waters engine. See Guy, 'North-eastern locomotive pioneers'.

p.95 The request for Hackworth to work on the Sabbath may have come from Hedley, as hinted at by Hackworth's biographer, his grandson, Robert Young (*Timothy Hackworth*, p. 76). It is conceivable that Hedley may have felt threatened by Hackworth: regardless of the claims made by Hedley's son (*Who invented the Locomotive Engine?*), it is conceivable that Hedley's contributions to the locomotive developments at Wylam may have been less than those of Hackworth. This speculation is supported by the apparent cessation of further locomotive developments at Wylam following Hackworth's departure.

p.95 The press announcement about Trevithick's arrival in Peru was given in *Government Gazette of Lima*, 12 February

1817, quoted in F. Trevithick, *Life of Richard Trevithick, with an Account of his Inventions* II (London: Spon, 1872), pp. 238–9.

p.95 The invoice for the first shipment of equipment to Lima is quoted in Trevithick, *Life of Richard Trevithick* II, p. 220.

p.96 The quote about Hackworth's 'great talents' is from John Hackworth's manuscripts, quoted in Young, *Timothy Hackworth*, p. 80.

p.96 According to Hackworth's grandson he managed the Stephensons' Forth Street locomotive works before attempting to go into business on his own, which seems to be confirmed by Hackworth's letter of introduction to the Stockton & Darlington Railway from Robert Stephenson & Co. (PRO, RAIL 667/1158).

p.97 The 'grievous injury' to *Hope* was reported in Richmond's *Local Records of Stockton*; sub-committee minutes, 5 October 1827, quoted in W. W. Tomlinson, *The North Eastern Railway: its Rise and Development* (Newcastle upon Tyne: Reid, 1914), p. 143.

p.99 Walker's quote about *Royal George* is from J. Walker, *Report to the Directors on the Comparative Merits of Locomotive and Fixed Engines as a Moving Power* (London: Liverpool & Manchester Railway, 1829), pp. 20–1 (PRO, ZLIB 4/57).

p.99 John Stephenson's quote about the finest locomotive in the world is given in Young, *Timothy Hackworth*, p. 163.

5 UP FROM THE MINE

p.103 The reference to 'Bob's engine-fire' is from S. Smiles, *The Life of George Stephenson, Railway Engineer* (third edition, London: John Murray, 1857), p. 5.

p.104 Bread prices are from *The Annual Register, or, A View of the History, Politics and Literature for the Year*, available on the Web at http://www.napoleon-series.org/research/abstract/economics/finance/c_bread1.html. When the Napoleonic Wars were over in 1815 the Corn Laws were enacted to protect farmers' profits by placing a duty on imported grain. This artificially prevented the price from falling below £4 a quarter (28 lb). Four pounds in 1815 would be worth almost £200 today, placing the price of 1 lb of flour at about £6.

p.105 Stephenson's quote about being a made man is given in Smiles, *Life of George Stephenson*, p. 13.

p.107 Stephenson's quote about perseverance is given in ibid., p. 494.

p.108 Stephenson's quote about fighting Nelson is from ibid., p. 27.

p.112 The reference to the 'considerable sum' of money George Stephenson had to pay to keep out of the militia is from ibid., p. 38.

p.112 George Stephenson's quote about being 'trusted in some small matters' is from ibid., p. 39.

p.113 Stephenson's conversation with Heppel about the High Pit engine is quoted in S. Smiles, *The Life of George Stephenson and of his son Robert Stephenson, comprising also a History of the Invention and Introduction of the Railway Locomotive* (New York: Harper, 1868), p. 132.

p.113 Dodds's conversation with Stephenson about the engineers being beaten is quoted in Smiles, *Life of George and Robert Stephenson*, p. 133.

p.115 Dodds's concerns over the modified engine knocking the house down are quoted in ibid., p. 133.

p.118 Stephenson's quote about Lord Ravensworth is from Smiles, *Life of George Stephenson*, p. 84.

p.121 The comment about forging cranks was made by Robert Stephenson in 1856 and quoted in Smiles, *Life of George and Robert Stephenson*, p. 167.

p.122 The patent of Stephenson and Dodds, with references to cranked axles and a chain connection between the front and back wheels, is discussed in Nicholas Wood's textbook on railways. In this book he figured a six-wheel locomotive with a chain connection (N. Wood, *A Practical Treatise on Rail-roads, and Interior Communication in General*, third edition, London: Longman, 1838, p. 292) but Robert Stephenson (quoted in Smiles, *Life of George and Robert Stephenson*, p. 166) says that 'four wheels . . . supported the engine' in the second locomotive. Wood also illustrated a four-wheel Killingworth locomotive (p. 294).

p.122 The comment about engine power being more than doubled was made by Robert Stephenson and quoted in Smiles (*Life of George and Robert Stephenson*), p. 168. Robert went on to attribute the invention of the 'steam-blast' to his father (p. 169), without any mention of Trevithick.

p.122 The reason for the strength of fish-belly rails derives from beam theory. When a beam, like a length of rail supported at either end on a stone block, is loaded with a weight the forces are not uniformly distributed. Through experiments, engineers determined that the forces were highest mid-way between the supports, which is why beams usually break in the middle rather than at either end. They also learned that beams were stronger when their greatest depth was in line with the weight being supported. A ruler held at either end can therefore support a much larger weight if it is held edge-wise-on than if it is flat-side-on. This explains why planks of timber used for floor joists are always placed on edge. It also explained to the engineers of Stephenson's day why fish-belly rails were the most economical way of using the cast iron, giving the maximum strength for the minimum amount of iron used.

p.124 Stephenson's priority over Davy's safety lamp has been supported by a recent reappraisal of the documented facts. See W. F. Watson, 'The invention of the miners' safety lamp: a reappraisal', *Transactions of the Newcomen Society* 70 (1998–99), pp. 135–41.

p.124 The quote from Stephenson's speech for delivery at the dinner held in his honour, and the reference to his saying he would never change, is given in J. C. Jeaffreson, *The Life of Robert Stephenson, FRS* I (London: Longman, 1866), pp. 40–1.

p.125 Pease's recollection of his meeting with Stephenson is quoted in A. E. Pease (ed.), *The Diaries of Edward Pease* (London, Hedley, 1907), p. 86.

p.126 Pease's quote about horse haulage on roads and rail is from ibid., p. 86.

p.127 James's prophetic statement about locomotives effecting a revolution is quoted in Smiles, *Life of George Stephenson*, p. 178.

p.127 James's declaration of Stephenson's genius is quoted in ibid., pp. 178–9.

p.128 The letter Pease wrote to his cousin Richardson on 10 October 1821, preserved in the Durham Record Office (D/Ho/C63/2), is quoted in part by A. Guy, 'North-eastern locomotive pioneers, 1805–27: a reassessment' in A. Guy and J. Rees (eds), *Early Railways: a Selection of Papers from the first International Early Railways Conference* (London: Newcomen Society, 2001), p. 131.

p.129 Pease's letter about Stephenson's appearances is quoted, in part, in Guy, 'North-eastern locomotive pioneers', p. 131. The letter was dated 26 January 1825.

p.130 Thomas Hindmarsh's claim about idle gossip is on the authority of Smiles, *Life of George and Robert Stephenson*, p. 215.

p.130 Stephenson's quote about soldiering is from a letter he wrote to his friend Joseph Cabry (sometimes spelt Cabery), 21 December 1819, quoted in full in W. O. Skeat, *George Stephenson: the Engineer and his Letters* (London: Institution of Mechanical Engineers, 1973), p. 36.

p.131 The quote about Trevithick's 'rude diving bell' and his speculation in pearl fishing is taken from a letter written by James Liddell to Francis Trevithick on 3 November 1869, given in Trevithick, *Life of Richard Trevithick* II, p. 249.

p.132 The quote about the prosperous state of the mines is from ibid. II, p. 246.

p.132 Trevithick's quote about the rich vein of copper is from ibid. II, p. 252.

p.132 Trevithick's quote about the arrival of the patriots is from ibid. II.

p.133 Trevithick's quote about Bolivar is from ibid.

p.133 Cochrane's disclosures on patronage revealed that seventy-six MPs held positions worth £150,000 (£6.5 million) per annum, while twenty-eight received annual pensions of £42,000 (£1.8 million).

6 FAMOUS SON OF A FAMOUS FATHER

p.135 The suggestion that Robert Stephenson's departure for South America was due to a rift with his father over William James was made by L. T. C. Rolt, *George and Robert Stephenson* (London: Longman, 1960; Penguin edition, 1984).

p.136 George Stephenson's letter about Robert's education was written to his friend William Locke on 31 March 1823, and is quoted in J. C. Jeaffreson, *The Life of Robert Stephenson, FRS* I (London: Longmans, 1866), pp. 59–60.

pp.136–137 Robert Stephenson's quotes about his experiences at Edinburgh University is contained in a letter he wrote to Michael Longridge in early December 1822, quoted in ibid. I, p. 58.

p.140 Robert Stephenson's letter to his father at the end of his Cornish expedition, written from Okehampton, 5 March 1824, is quoted in ibid. I, p. 69.

pp.140–141 Robert Stephenson's appeal for his father's support is cited in ibid. I, p. 70.

p.141 George Stephenson's comment to Longridge about Robert's good spirits is contained in a letter he wrote from Liverpool on 15 June 1824, cited in ibid. I, pp. 75–6.

pp.141–142 Longridge's letter to Robert Stephenson expressing his regret at his long absence is in the library of the Institution of Mechanical Engineers (IMS 164).

p.143 Stephenson's words to the manager at Bogotá about the indolent miners are contained in a letter, quoted in Jeaffreson, *Life of Robert Stephenson* I, pp. 87–8.

p.143 Stephenson's words to the drunken miners are quoted in ibid. I, p. 90.

p.144 George Stephenson's postscript about armed opposition to the survey is in a letter to Joseph Pease, written 19 October 1824, quoted in W. O. Skeat, *George Stephenson: the Engineer and his Letters* (London: Institution of Mechanical Engineers, 1973), pp. 76–7.

p.145 Locke's letter to Robert Stephenson about the opening of the Stockton & Darlington Railway, written 24 November 1825, is at the Institution of Mechanical Engineers (IMF 165).

p.146 John Rennie's reply to the directors referring to his brother's return to London is paraphrased in the minutes of the meeting of 5 June 1826 (PRO, RAIL 371/1/1, p. 14).

p.146 The quote regarding the conditions under which the Rennies would accept the superintendence of the railway is taken from the minutes of the meeting of 9 June 1826 (ibid., p. 15).

p.146 The quote about Rennie refusing to work with Stephenson is from the minutes of the meeting of 17 June 1826 (ibid., intervening page between pp. 15 and 16).

p.147 Vignoles's comments on his strained relationship with George Stephenson are contained in a draft letter quoted in K. H. Vignoles, *Charles Blacker Vignoles, Romantic Engineer* (Cambridge: Cambridge University Press, 1982).

p.147 For more information about Joseph Locke see N. W. Webster, *Joseph Locke: Railway Revolutionary* (London: Allen & Unwin, 1970).

p.148 That *Hope* could not initially be made to work is discussed in R. Young, *Timothy Hackworth and the Locomotive* (Shildon: 'Stockton & Darlington Railway' Jubilee Committee, 1975; first published 1923), p. 140.

p.148 Longridge's hope of 'Robert's early return to England . . .' was contained in a letter he wrote to Richardson on 7 March 1825, quoted in J. G. H. Warren, *A Century of Locomotive Building by Robert Stephenson & Co., 1823–1923* (Newcastle upon Tyne: Reid, 1923), p. 65.

p.148 Locke's letter about business at Forth Street being 'not so briskly as it has done' was written to Robert Stephenson on 25 February 1827 and is at the Institution of Mechanical Engineers (IMS 165).

p.149 Nicholas Wood's comments on locomotive speeds, apparently written in 1825, were published in *Mechanics' Magazine* 323 (17 October 1829), p. 134.

p.150 Pease's reminiscences about the financial woes of the Stockton & Darlington Railway are contained in a letter he wrote in 1857, cited in E.M.S.P. [Mrs Paine], *The two Jameses and the two Stephensons* (1861, reprinted Dawlish: David & Charles, 1961), p. 47.

p.150 Stephenson's last South American letter to Longridge is quoted in Jeaffreson, *Life of Robert Stephenson* I, pp. 101–4.

p.151 Stephenson's recollection of his meeting with Trevithick in Cartagena is given in F. Trevithick, *Life of Richard Trevithick, with an Account of his Inventions* II (London: Spon, 1872), p. 274.

pp.151–152 Fairbairn's recollection of the meeting between Trevithick and Stephenson, and Hall's corroboration, are given in Trevithick, *Life of Richard Trevithick* II, pp. 272–3. Hall is said to have been 'an officer in the Venezuelan and the Peruvian services'.

pp.152–153 An account of Trevithick's expedition across the Cordilleras is given in ibid. II, p. 269.

p.153 Trevithick's last adventure, in Venezuela, is reported in ibid. II, p. 272.

p.153 The quote about the 'good passage home' is from a letter Trevithick wrote to Gerard on 15 November 1827, quoted in full in ibid. II, pp. 277–8.

p.154 Stephenson's account of the early part of his voyage home, including the cannibalism, is contained in a letter to Mr Illingsworth, written 1 March 1828. The letter is quoted in part in Jeaffreson, *Life of Robert Stephenson* I, p. 106, and also in S. Smiles, *The Life of George Stephenson and of his son Robert Stephenson, comprising also a History of the Invention and Introduction of the Railway Locomotive* (New York: Harper, 1868), pp. 308–9.

p.155 Jeaffreson's reference to 'a mass of evidence' is given on p. vi of his preface to *The Life of Robert Stephenson.*

p.155 Jeaffreson's comment about Trevithick's impression on Stephenson is given in *The Life of Robert Stephenson*, p. 105.

p.155 Stephenson's letter to Longridge about improving locomotives, written 1 January 1828 from Liverpool, is quoted in ibid., pp. 114–15.

7 ROCKET ON TRIAL

pp.157–158 Extracts from *The Times* are for Thursday 8 October 1829; notice of Fanny Kemble's theatre appearance was given in *The Times* on Tuesday 6 October 1829.

p.159 The extract from the *Liverpool Chronicle* is from Saturday 10 October 1829, p. 328.

p.159 The quotes about 'full compliment of water' and 'fuel in the fireplace' are from J. U. Rastrick, N. Woods and J. Kennedy, 'Report of the Competition on the Liverpool & Manchester Railway, to the Directors of the Liverpool & Manchester Railway' (1829), transcript from the Science Museum, London, Science and Society Picture Library, NRM/RH1/B460201B–B460210B.

p.161 For a useful discussion of the use of springs in safety valves see R. Gibbon, 'Rings, springs, string and things: the national collection pre-1840', in A. Guy and J. Rees (eds), *Early Railways: a Selection of Papers from the first International Early Railways Conference* (London: Newcomen Society, 2001), pp. 208–18.

p.165 George Stephenson's comment about the 'complicated job' is from a letter to Robert, written 15 April 1828, quoted in J. C. Jeaffreson, *The Life of Robert Stephenson, FRS* I (London: Longman, 1866), p. 120.

p.168 That *Rocket*'s original valves have not been preserved is given by Bailey and Glithero in their superbly informative and comprehensive book: M. R. Bailey and J. P. Glithero, *The Engineering and History of* Rocket (London: Science Museum, 2000).

pp.168–169 The timings and other data for *Rocket*'s trial are taken from Rastrick *et al.*, 'Report of the Competition'.

p.169 Robert Stephenson's quote about changing gears is contained in a letter he wrote to Booth on 21 August 1829 (PRO, RAIL 1088/88).

pp.174–175 Booth's recollection of his scheme, quoted in Bailey and Glithero, *The Engineering and History*, p. 14, is from Alfred Booth, *Some Memories, Letters and other Family Records, written and arranged by his Daughter, Harriet Anna Whitting* (printed for private circulation, Liverpool, 1917).

A copy of this rare publication is available in the Liverpool Record Office.

p.175 The letter to Trevithick about a multi-flue boiler is undated but the year 1816 was later added (Science Museum, MS 347).

p.175 Robert Stephenson's quote about the tubes becoming furred is from a letter he wrote to Samuel Smiles, who was at that time writing his father's biography. The letter is quoted in Jeaffreson, *Life of Robert Stephenson* I, pp. 126–7.

p.176 The consequences of the relationship between size and surface area are discussed at great length in C. McGowan, *Diatoms to Dinosaurs: The Size and Scale in Living Things* (Washington DC: Island Press; London: Penguin Books, 1994).

p.176 Booth's comment on the advantages of multiple tubes is quoted from a letter he wrote to Smiles, published in S. Smiles, *The Life of George Stephenson and of his son Robert Stephenson, comprising also a History of the Invention and Introduction of the Railway Locomotive* (New York: Harper, 1868), p. 318.

p.176 For a good account of Séguin and his multi-tubular boiler see J. G. H. Warren, *A Century of Locomotive Building by Robert Stephenson & Co., Newcastle-upon-Tyne* (Newcastle upon Tyne: Reid, 1923), pp. 135–9.

p.176 Robert Stephenson's attribution of *Rocket* to his father is from a letter he wrote to Smiles, quoted in Jeaffreson, *Life of Robert Stephenson* I, pp. 126–7.

p.177 George Phipps's recollection of Robert Stephenson and the building of the *Rocket* is from a letter he wrote to *The Engineer*, 17 September 1860, p. 217.

p.178 Stephenson's call for the assistance of 'the oracle' is quoted from Jeaffreson, *Life of Robert Stephenson* I, p. 139.

p.179 Phipps's contradiction of Smiles's account of how the boiler tubes were fitted in *Rocket* is given in *The Engineer*, 24 September 1860, p. 230.

pp.179–180 The reference to the letter Robert Stephenson wrote to Henry Booth on 3 August 1829 is PRO, RAIL 1088/88 (part of a folder containing several letters).

p.182 Part of the expense of the *Rocket*'s firebox was the high cost of copper. When *Rocket* underwent repairs following an accident at Chat Moss – a marshy area that had given George Stephenson considerable trouble during the construction of the line – the solid rear plate of the firebox was replaced with one consisting of a double-skinned water jacket. This was made of wrought iron, which probably explains why it has survived, while the valuable copper saddle has long since disappeared.

p.183 The experiments on the deleterious effects of coupling rods were reported in a paper by R. Gibbon and R. Lamb entitled 'Running with your Breeks down? The Resistance of Coupled Locomotive Wheels', given at the second International Early Railway Conference, Manchester, 6–19 September 2001.

p.184 Robert Stephenson's second progress letter to Booth, after a delay of almost three weeks, was written on 21 August 1829 (PRO, RAIL 1008/88).

pp.185–186 Robert Stephenson's letter to Booth about the need to fit stays was written on 26 August 1829 (PRO, RAIL 1008/88).

pp.186–187 Robert Stephenson's quote about Burstall's spying is from the letter he wrote to Booth on 31 August 1829 (PRO, RAIL 1008/88).

p.187 Robert Stephenson's quote about the wheels is from Ibid.

pp.187–188 The reference for Robert Stephenson's letter to Booth of 5 September 1829, quoted here in its entirety, is PRO, RAIL 1008/88.

p.189 The account of *Rocket*'s back wheels not fitting when they arrived at Rainhill was from a letter written by Robert Stannard, the lad who was given the ride at the end of the first day of the trials. The letter was written more than half a century after the event, and was published in *The Engineer* 58 (17 October 1884), p. 303.

p.190 The quote about the postponement of *Novelty*'s trial until Monday is from Rastrick *et al.*, 'Report of the Competition'.

8 THE PEOPLE'S CHOICE

pp.192–193 Reference to *Novelty*'s 'air-compressing apparatus' is made in the *Mechanics' Magazine* 323, Saturday 17 October 1829, p. 142. It is referred to again in the subsequent issue (324, Saturday 24 October, p. 150) as 'the bellows-sort of apparatus'.

p.194 Details of Vaughn's patent blower and its adoption in the original *Novelty* are to be given in a paper by Richard Lamb and Richard Gibbon, entitled 'Blowing up *Novelty*', at the third International Early Railway Conference, York, England.

p.195 The quote about the visit of Braithwaite 'in company with his friend' is from J. U. Rastrick, N. Woods and J. Kennedy, 'Report of the Competition on the Liverpool & Manchester Railway, to the Directors of the Liverpool & Manchester Railway' (1829), transcript from the Science Museum, London, Science and Society Picture Library, NRM/RH1/B460201B–B460210B.

p.195 The quote about Braithwaite's declaration 'that his Engine would be all complete' is from ibid.

p.195 The judges' response is from ibid.

p.197 The quote about *Novelty*'s burst feed pipe is from ibid.

p.197 Rastrick's quote about the reason for *Novelty*'s burst pipe is taken from a transcript appended to C. F. Dendy Marshall, *Centenary History of the Liverpool & Manchester Railway* (London: Locomotive Publishing Co., 1930), p. 176.

p.198 The account of *Rocket*'s singular display is from *The Times*, Monday 12 October 1829.

p.199 Vignoles's account of *Novelty*'s demonstration for 'about forty-five ladies and gentlemen' is given in *The Albion*, 12 October 1829, p. 325; *Mechanics' Magazine* 323 (Saturday 17 October 1829), pp. 138–9.

p.199 The report about Dr Traill is from the *Liverpool Chronicle*, Saturday 17 October, p. 335.

p.200 That Hackworth had helped Braithwaite and Ericsson get *Novelty* ready to compete is said by R. Young (*Timothy Hackworth and the Locomotive*, Shildon: 'Stockton & Darlington Railway' Jubilee Committee, 1975; first published 1923, p. 195) to have been acknowledged by Ericsson.

p.200 The quote about Hackworth not requiring any further time is from Rastrick *et al.*, 'Report of the Competition'.

9 THE DARK HORSE

p.201 The quote about 'some novelty' being witnessed was from *The Liverpool Chronicle* for Saturday 17 October; p. 335.

pp.202–203 A transcription of 'Rastrick's Rainhill Notebook', together with reproductions of his sketches, was given in the appendix of C. F. Dendy Marshall, *Centenary History of the Liverpool & Manchester Railway* (London: Locomotive Publishing Co., 1930), p. 172. *Sans Pareil* as she stands today is a little different from the way she was at Rainhill. Aside from having different wheels, the furnace end of the flue tube appears to have been modified. Instead of the extension of the boiler casing being restricted to the upper half of the flue tube, it completely sheaths it.

p.204 The quote from the judges' report is from J. U. Rastrick, N. Wood and J. Kennedy, 'Report of the Competition on the Liverpool & Manchester Railway, to the Directors of the Liverpool & Manchester Railway' (1829), transcript from the Science Museum, London, Science and Society Picture Library, NRM/RH1/B460201B–B460210B.

p.205 Hackworth's grandson, Robert Young, was quite sure William Gowland was one of *Sans Pareil*'s crew at Rainhill (*Timothy Hackworth and the Locomotive*, Shildon: 'Stockton & Darlington Railway' Jubilee Committee, 1975, first published 1923, p. 304). However, according to Rolt (*George and Robert Stephenson: the Railway Revolution* (London: Longman, 1960, p. 5), it was Thomas George.

p.205 The quote about 'a very powerful competitor', was from *Mechanics' Magazine* 323, Saturday 17 October 1829, p. 138.

p.206 John Dixon's disparaging remark about *Sans Pareil* is quoted from a letter he wrote to his brother James on 16 October 1829 (Science Museum Library, MS 1573).

p.207 The quote about *Sans Pareil*'s draft being so great as to eject fuel from the chimney is from a letter written by John Dewrance, who was at the trials: *The Engineer* 4 (21 August 1857), p. 134.

p.207 Robert Stephenson's quote about *Sans Pareil*'s draught being so great as to eject fuel from the chimney is from S. Smiles, *Lives of the Engineers* III (1862), quoted in J. G. H. Warren, *A Century of Locomotive Building by Robert Stephenson & Co., 1823–1923* (Newcastle upon Tyne: Reid, 1923; reprinted, Newton Abbot: David & Charles, 1970), pp. 226–227.

p.208 The circumstantial evidence that *Rocket* may have had a sight gauge, given by Bailey and Glithero (M. R. Bailey and J. P. Glithero, *The Engineering and History of* Rocket, London: Science Museum, 2000, p. 85), is based on the fact that *Northumbrian*, built shortly after her, was so equipped. This is from the personal recollection of Fanny Kemble, following her footplate ride in August 1830 (see Chapter 11).

p.213 The quote about *Sans Pareil*'s fusible plug melting near the grandstand is from Rastrick *et al.*, 'Report of the Competition'.

p.213 The quote from Rastrick's notebook is taken from a transcription of 'Rastrick's Rainhill Notebook', given in the appendix of Dendy Marshall, *Centenary History*, p. 172.

p.213 The judges' additional information regarding *Sans Pareil*'s breakdown is from Rastrick *et al.*, 'Report of the Competition'.

p.213 The quote from *The Times* about *Sans Pareil*'s performance is from 16 October 1829. The identical report in *The Albion* appeared on 19 October 1829.

p.214 The account of the incident where a man fell on the track as *Sans Pareil* was approaching is from *The Albion* of 12 October 1829, p. 325.

p.214 The quote regarding *Sans Pareil* having another trial is from the *Mechanics' Magazine* 323 (Saturday 17 October 1829), p. 140.

p.214 Dixon's comment about 'Malt Sprouts' is from his letter of 16 October 1829 (Science Museum Library, MS 1573).

p.215 The values given for the heat-exchange areas of *Sans Pareil* and *Rocket* are from the second edition of Nicholas Wood's book. The *Rocket* estimate is probably accurate, because the external surface area of each tube is easily calculated. However, it is difficult to estimate the area of the external surface of *San Pareil*'s flue tube, so that value is less reliable.

10 WINNING DAY

p.217 The account of the initial damage to *Novelty*'s boiler is given in the *Mechanics' Magazine* 324 (Saturday 24 October 1829), pp. 150–1.

p.218 The quote about raising steam in 'less than 40 minutes' is from the *Mechanics' Magazine* 323 (Saturday 17 October 1829), p. 140.

p.219 The judges' quote about the failure of *Novelty*'s boiler is from J. U. Rastrick, N. Woods and J. Kennedy, 'Report of the Competition on the Liverpool & Manchester Railway, to the Directors of the Liverpool & Manchester Railway' (1829), transcript from the Science Museum, London, Science and Society Picture Library, NRM/RH1/B460201B-B460210B. That of Stephenson and Locke is from R. Stephenson and J. Locke, *Observations on the Comparative Merits of Locomotive & Fixed Engines as applied to Railways . . . with an Account of the Locomotive Engines at Rainhill, Liverpool*, printed by Wales and Baines, 1829 (PRO, ZLIB 4/57, p. 78).

pp.219–220 The judges' reasons for denying Hackworth a second trial are quoted from Rastrick *et al.*, 'Report of the Competition'.

p.220 The locomotive *Samson*, on public display in Canada at the Museum of Industry, Stellarton, Nova Scotia, was the subject of an intensive investigation. See M. R. Bailey and J. P. Glithero, 'Learning through restoration: the *Samson* locomotive project' in A. Guy and J. Rees (eds), *Early Railways: a Selection of Papers from the first International*

Early Railways Conference (London: Newcomen Society, 2001), pp. 278–93.

p.221 Dixon's comments on *Sans Pareil*, on *Perseverence*, and on their respective builders, are quoted from a letter he wrote to his brother James on 16 October 1829 (Science Museum Library, MS 1573).

p.222 The quotes from *The Liverpool Mercury* about *Novelty* and about Stephenson are given in the *Mechanics' Magazine* 324 (Saturday 24 October 1829), p. 146.

pp.222–223 The quotes from the *Mechanics' Magazine* are from Saturday 24 October 1829, pp. 146–8.

p.223 The quotes from the *Mechanics' Magazine* about the history of the Liverpool & Manchester Railway are from ibid.

p.224 The locomotive *Samson*, tested at the same time as *William IV* and *Queen Adelaide*, is not to be confused with the Hackworth engine of the same name that went to Nova Scotia.

p.225 The quote from the *Manchester Guardian* article about *William IV*, published 26 February 1831, is from the *Mechanics' Magazine* 397 (1831), p. 37.

pp.225–226 The quote from the *Manchester Guardian* of 26 February 1831 about 'Mr Stephenson and his engines' is from ibid.

p.226 The *Mechanics' Magazine*'s rebuttal appeared in ibid.

p.226 The quote from Braithwaite and Ericsson's letter to the directors is taken from J. G. H. Warren, *A Century of Locomotive Building by Robert Stephenson & Co., 1823–1923 823–1923* (Newcastle upon Tyne: Reid, 1923), p. 82.

p.226 Ericsson's quote about the engines being 'utter failures' is from W. C. Church, *The Life of John Ericsson* I (New York; Scribner, 1906), p. 62.

p.227 Kitching's letter to Hackworth is quoted, in part, in R. Young, *Timothy Hackworth and the Locomotive* (Shildon: 'Stockton & Darlington Railway' Jubilee Committee, 1975; first published 1923), p. 199.

pp.228–229 Hackworth's letter to the directors, probably written some time in November 1829, is in the Library of the Science Museum (MS 1562, typed transcript of original letter).

11 TRIUMPH AND TRAGEDY

p.231 The recollection of Jenyns is from R. F. Vaughan, *Events in the Life of the Rev. Leonard Jenyns in the year 1830*, http:// www.rogerco.freeserve.co.uk.

p.234 Rotten boroughs were those with very few voters, sometimes only dozens. Such boroughs not only skewed the representation in the House of Commons but were also highly prone to voting irregularities, as when landowners pressured their employees to vote for the candidate of his choice.

pp.234–235 Creevey's experience in riding a train is from the Creevey papers and is quoted in L. T. C. Rolt, *George and Robert Stephenson: the Railway Revolution* (London: Longman, 1960; reprinted London: Penguin Books, 1984), pp. 189–90.

p.235 The quotes about the directors' tour of inspection are from the minutes of the directors' meetings (PRO, RAIL 371/42, part 1).

pp.236, 238, 241, 242, 243, 244 Fanny Kemble's account of her first locomotive journey is contained in a letter written from Liverpool on 26 August 1830 which is part of her memoirs, *Record of a Girlhood* II (1878), quoted in part in J. G. H. Warren, *A Century of Locomotive Building by Robert Stephenson & Co., 1823–1923* (Newcastle upon Tyne: Reid, 1923), pp. 246–9.

p.242 George Stephenson's recollections of Chat Moss are quoted from S. Smiles, *The Life of George Stephenson and of his son Robert Stephenson, comprising also a History of the Invention and Introduction of the Railway Locomotive* (New York: Harper, 1868), p. 287.

p.244 The quote from the report of Huskisson's Liverpool speech is from *The Liverpool Courier*, 15 September 1830, p. 296.

p.245 The quote about the 'assemblage of rank' is from *The Liverpool Mercury*, 16 September 1830, p. 304.

p.245 The quotes about 'all classes of the community', 'the lower orders' and 'absorbed in the anticipation' are all from *The Liverpool Courier*, 22 September 1830, p. 298.

p.248 The quote about 'arrow-like swiftness' is from *The Liverpool Mercury*, 17 September 1830, p. 303.

p.249 The quote from *The Liverpool Courier* about uniforms is from 22 September 1830, p. 298.

pp.249–250 The quotes about 'vehicles of every description' and about crossing Sankey viaduct are from ibid.

p.250 There is some variation in the accounts of how many trains passed the stationary *Northumbrian* prior to the accident. According to the first correspondent of *The Times* for 17 September 1830, who was aboard *Phoenix*, only the latter had passed the Duke's train. This is confirmed by their second correspondent, who stated *Rocket* was the second locomotive upon the left-hand track. However, *The Liverpool Courier* (22 September 1830, p. 208) stated that 'Two trains had passed the duke's . . .', while *The Liverpool Mercury* (16 September 1830, p. 304) stated that 'After three of the engines with their trains had passed the Duke's carriage . . .' The two corresponding accounts in *The Times* are more persuasive than the conflicting accounts of the other two sources.

p.250 The quote about 'joyous excitement' is from *The Liverpool Courier*, 22 September 1830, p. 298.

p.250 Huskisson's quote to Sandars is from *The Times*, 17 September 1830.

p.250 The quote about the two statesmen is from *The Liverpool Courier*, 22 September 1830, p. 298.

p.251 The quote, 'Get in . . .' is from ibid.

p.252 The quote pertaining to Holmes is from ibid.

pp.252–253 The quote describing Huskisson's injuries is from *The Times*, 17 September 1830.

p.253 Huskisson's comment about his pending death is from *The Liverpool Courier*, 22 September 1830, p. 298.

p.253 Huskisson's cool manner is quoted in *The Times*, 17 September 1830.

p.254 The quote about Huskisson being 'pale and ghastly', and that of his wife in an 'agony of tears', are from ibid.

p.254 That it was the reeves of Manchester and Salford who persuaded the Duke is from ibid. The Duke's remark is quoted from *The Liverpool Courier*, 22 September 1830, p. 298.

p.256 The reference to the 'lower orders', a common term for working-class people at the time, is from ibid.

p.257 The quote about 'the cheerings of the great body . . .' is from ibid.

p.257 The quote about 'a poor and wretchedly dressed man' is from ibid.

p.257 The quote about the 'cold collation' is from ibid.

p.257 The quote about the sumptuous wines is from *The Times*, 17 September 1830.

p.257 The quote about the Duke shaking hands with 'thousands of persons' is from *The Liverpool Courier*, 22 September 1830, p. 298.

p.258 The quote about Mr Lavender's concerns is from ibid.

p.259 The quote about plunging the party into deeper distress is from *The Liverpool Mercury*, 16 September 1830, p. 304.

p.260 The quote about the 'painful sensation' is from *The Liverpool Courier*, 22 September 1830, p. 298.

p.261 The quote about 'drunken fellows' is from ibid., p. 299.

12 BOOM AND BUST

pp.262–265 and revenue data for the early years of the Liverpool & Manchester Railway, and comparative costs for stagecoach travel, are from F. Ferneyhough, *Liverpool & Manchester Railway, 1830–1980* (London: Book Club Associates, 1980).

p.266 Stephenson's account of the dangers of travelling by road is quoted from L. T. C. Rolt, *George and Robert Stephenson: the Railway Revolution* (London: Longman, 1960; reprinted London: Penguin Books, 1984), p. 220.

p.266 Data for mileage of line operational from H. Pollins, *Britain's Railways: an Industrial History* (Newton Abbot: David & Charles, 1971). Data for the number of new railways sanctioned by Parliament are from Pollins, *Britain's Railways*, and from U. Hielscher, *The Emergence of the Railway in Britain* (London: International Bond & Share Society, 2001).

p.266 Employment data from M. C. Reed, *Railways in the Victorian Economy* (Newton Abbot, David & Charles, 1978).

p.266 The data for railway shares are from Hielscher, *Emergence of the Railway*, and Pollins, *Britain's Railways*.

p.268 Stephenson's letter to Longridge about Hudson was written 22 November 1845, and is quoted in full in W. O. Skeat, *George Stephenson: the Engineer and his Letters* (London: Institution of Mechanical Engineers, 1973), pp. 224–45.

pp.268–269 The verse from *Illustrated London News*, published in 1845, is quoted from Ferneyhough, *Liverpool & Manchester Railway*, p. 139.

p.269 The quote from Brunel about the railway world being mad is from ibid.

p.269 Stephenson's prediction is from a letter to his solicitor, quoted in Rolt, *George and Robert Stephenson*, p. 292.

13 BEGINNINGS AND ENDINGS

p.272 The quote about Trevithick's 'wholly inapplicable' gun carriage is from a letter, dated 21 February 1828, he received from the Office of Ordnance, quoted in full in F. Trevithick, *Life of Richard Trevithick, with an Account of his Inventions* II (London: Spon, 1872), p. 287.

p.273 Details of Trevithick's funeral, which have often been misreported, are given in Anthony Burton's thoroughly readable book *Richard Trevithick: Giant of Steam* (London: Aurum Press, 2000).

pp.273–274 The quote from Trevithick's letter to Gilbert is taken from Trevithick, *Life of Richard Trevithick* II, pp. 395–6.

p.276 The quote 'beware of the bubbles' is from S. Smiles, *The Life of George Stephenson and of his son Robert Stephenson, comprising also a History of the Invention and Introduction of the Railway Locomotive* (New York: Harper, 1868), p. 350.

p.277 The quotes pertaining to Sir Astley Cooper and Stephenson's comment are from ibid., pp. 350–1.

p.282 Arnold's quote about the tunnel is from L. T. C. Rolt (*George and Robert Stephenson: the Railway Revolution*, London: Longman, 1960; reprinted London: Penguin Books, 1984), p. 230.

p.282 The term 'navvy' was used for a labourer, originally for those working on a navigation or canal.

p.282 The cost estimates for Kilsby tunnel are from D. Beckett, *Stephensons' Britain* (Newton Abbot: David & Charles, 1984), p. 84, while those for the entire railway are from Rolt, *George and Robert Stephenson*, p. 245.

p.283 Stephenson's comment about his reputation breaking 'like an eggshell' is quoted from Rolt, *George and Robert Stephenson*, p. 231.

p.287 The question of the solar alignment of the Box tunnel is discussed in C. P. Atkins, 'Box railway tunnel and I. K. Brunel's birthday: a theoretical investigation', *Journal of the Astronomical Society* 95, 6 (1985), pp. 260–2.

p.289 The comment about the Stephenson long-boiler locomotive derailing was made by Daniel Gooch, designer of *Ixion* and so many other broad-gauge locomotives, and quoted in J. G. H. Warren, *A Century of Locomotive Building by Robert Stephenson & Co., 1823–1923* (Newcastle upon Tyne: Reid, 1923), p. 385.

pp.290–291 The quote about the formation of coal is from E. O. Gordon, *The Life and Correspondence of William Buckland, DD, FRS* (London: John Murray, 1894) pp. 167–168.

pp.291–292 Stephenson's experience of doctors is contained in a letter to Longridge, written 22 November 1845, quoted in full in W. O. Skeat, *George Stephenson: the Engineer and his Letters* (London: Institution of Mechanical Engineers, 1973), pp. 224–5.

p.292 The quote about success and immediate prosperity is from W. C. church, *The Life of John Ericsson* I (New York, Scribner's, 1906), p.85

p.293 Ericsson's depth-sounding device was built in co-operation with F. B. Ogden, whom he duly acknowledged. It appears that Ericsson developed Ogden's original idea.

p.293 The quote from *The Times* about steam navigation is from Church, *Life of John Ericsson* I, p. 94.

p.295 Quotes from Ericsson's letter to Lincoln, written 29 August 1861, are from ibid. I, p. 246.

p.295 The reference to 'Ericsson's Folly' is from ibid., p. 255

pp.295–296 Ericsson's rant about *Monitor*'s rudder is from ibid., p. 256.

p.296 The quotation about seaworthiness is from a letter Stimers wrote to Ericsson aboard *Monitor* on 9 March 1862, quoted in full in ibid. I, pp. 280–1.

p.296 The quote about *Monitor* sliding away from *Merrimac*'s stern, and the quote from Catesby Jones, are both taken from an account of the battle by a Confederate soldier, given in ibid. I, pp. 284–6.

p.296 The quote by *Monitor*'s chief engineer congratulating Ericsson on his 'great success' is from ibid. I, p. 280.

p.297 The quote from the joint resolution of Congress is from ibid. I, p. 294.

p.297 Data for the cost of *Monitor* are from ibid. I, p. 269.

p.297 Data for the additional monitors are from ibid. I, p. 8.

p.298 Ericsson's conversation with Dr Markoe is taken from ibid. I, p. 322.

p.300 Hackworth paid the Stockton & Darlington Railway an annual fee for the use of its equipment, calculated at 5 per cent of its capital cost. This amounted to an annual cost of about £1,000 (£64,000). In return he received 0.4*d* per ton per mile for hauling freight.

p.301 Hackworth supplied the Albion coal mines of Nova Scotia with three locomotives, *Samson*, *Hercules* and *John Buddle*.

p.303 John Hackworth's letter of challenge to Robert Stephenson is quoted in full in R. Young, *Timothy Hackworth and the Locomotive* (Shildon: 'Stockton & Darlington Railway' Jubilee Committee, 1975; first published 1923), p. 328.

p.304 Robert Stephenson's quote about shrinking from responsibility is taken from a much longer quotation given by Smiles, *Life of George and Robert Stephenson*, p. 444.

p.304 Stephenson's rejection of a suspension bridge across the Menai Strait is from ibid., pp. 442–3.

p.310 Robert Stephenson's quote about the tubes is from ibid., p. 454.

p.312 The quote about falling in the mud is from ibid., p. 486.

p.312 Robert Stephenson's conversation about his father's desk is from J. C. Jeaffreson, *The Life of Robert Stephenson, FRS* I (London: Longman, 1866), p. 239.

p.312 The quote about Robert's meeting with an old friend is from ibid., p. 240.

p.313 The quote about the trip to Scotland aboard *Titania* is from a letter by Mr Kell, one of Stephenson's friends who was on the trip, written 26 October 1857. The purported depth of 170 fathoms is incorrect, because the deepest basins in Loch Ness are no more than about 600 ft. The letter is quoted in full in Jeaffreson, *Life of Robert Stephenson*, pp. 241–3.

p.314 Stephenson's comments about the climate and his health are quoted from a letter written from Alexandria, 22 December 1856, quoted in full in ibid., pp. 246–7.

p.314 The quote about Brunel as carefree as a schoolboy is from ibid., p. 249.

p.314 The quotes about Stephenson's health after his return from Cairo are from ibid., p. 249.

p.315 The quote about 'dropsy' is from ibid., p. 260.

p.315 The quote of his death striking the feelings of the country are from ibid., p. 260.

p.316 Stephenson's quote about railways, and about his father, is from his address to the Institution of Civil Engineers, in S. Smiles, *The Life of George Stephenson, Railway Engineer* (third edition, London: John Murray, 1857), pp. 515 and 545.

Index

Admiralty
buys a Trevithick engine 53
rejects Trevithick's gun carriage 272
amalgam 132
Anglican Church 91
Arnold, Thomas, famous headmaster 281–2
assigned load, of competing locomotives 30
atmospheric engines
demonstration of principles 39
distinct from true steam engines 37
function of beam 39
George Stephenson's familiarity 106
illustrated 38
less effective at high altitudes 60
originally for pumping water 40
reciprocating motion 40
working cycle explained 39–40
working principles 37, 39
atmospheric pressure 39
BBC television *Timewatch* 193
beam engine, alternative term for atmospheric engine 37
Bedlington Iron Works
boiler plates varied in thickness 82
built *Sans Pareil's* boiler 81
Longridge associated with 82
Black Callerton, village near Dolly Pit mine 107
Blackett, Christopher
employees concerned for jobs over locomotives 92
first locomotive not very successful 91
hires Thomas Waters to build a locomotive 90
impressed with Hackworth 86
not an engineer 87
owner of Wylam Colliery 57
refuses delivery of Gateshead locomotive 57
renews interest in locomotives 86
stops locomotive visits to Wylam 94
blast furnace
coke fired, reduced cost of iron 87
details of process 327–8
blast pipe 56
Blenkinsop locomotive 89
cogwheel drive 88
first commercially successful railway 90
high wear and tear 90
other collieries adopt system 90
public demonstration 90
Blenkinsop, John
built *Salamanca* 35
claims for his locomotive 90
cogwheel-drive locomotive 88
former viewer at Middleton Colliery 87
locomotive pioneer 35
blood transfusions 256
blood-letting 291
Bob's engine-fire, popular children's rendezvous 103
body-snatchers 273
boiler barrel 12
boilermaker, skilled job 81
boilers
barrel 12
barrels rarely exploded 36

dangers of low water levels 212
explosions usually involved flue tube 36
flue types 13
heat circulation problems 175, 215
hydraulic pressure testing 185
multi-tubular 175
multi-tubular type not size-constrained 220
multiple flue designs 175
problems building *Rocket*'s 178
replenishing water 69
return-flue 13
size-constraint of flue-tube type 220
straight-through flue 13
uneven heating of flue tube 36
Bolivar, Simon
arrives in Lima 133
conscripts Trevithick into his army 133
rides into Ecuador 132
sends Trevithick on mission to Bogotá 134
Bolton and Leigh Railway 229
Booth
company treasurer 7
explains multi-tubular boiler 176
helps fund *Rocket* 15
idea for revolutionary new boiler with multiple tubes 174–5
initial idea for *Lancashire Witch*'s boiler 164
learns of *Rocket*'s test run from Robert 187–8
proposes competing in trials with George Stephenson 174–5
received fanciful proposals for trials 7
Rocket entry with George and Robert Stephenson 100
vital dates 7
Boulton & Watt
attempt to banish high-pressure steam engines 54
cannot help Uville 60
charge user fees 42
employ William Murdock 45
legal dealings with Trevithick 43
make capital of Trevithick's explosion 53
partnership formed 42
patent expiry 50, 51
sue patent violators 43
Boulton, Matthew, talented engineer and businessman 42
Box tunnel, Brunel birthday legend 287
Braithwaite 63
age at Rainhill 10
boyish appearance 62
co-builder of *Novelty* 10
helped father salvage wreck 72
insists *Novelty* ready for 2nd trial, against judges' counsel 195
kind and considerate 65
life after Rainhill 298–9
owned engineering works 23, 72
predeceases Ericsson 298
sudden death 299
vital dates 10
Braithwaite and Ericsson
fire engine 74–5
given time to repair *Novelty*'s bellows 76
gratitude to L & M Railway 226
good friends 70
stayed out of the fray 226
realize cause of *Novelty* failure during first trial 76
unready to compete first day 24
brakesman 107
Brandreth
compensates wounded crew man 35
designed *Cycloped* 26
former director L & M Railway 26
gives second demonstration of *Cycloped* 77
takes credit for dandy cart 28
bread prices 232
bread riots 104
Bright's disease 297, 314

Britain
better opportunities than on Continent 72
entrepreneurial environment 72
growth of consumer society 73
laissez-faire policy of Government 73
landowners free to exploit their mineral wealth 72
led the industrial world 72
Britannia bridge 305
floating first girder into position 309
raising the girders 310
Robert Stephenson give tour on *Titania* trip 313
spanning Menai Strait 308
broad gauge 288
Brunel, Isambard Kingdom 286
Bright's disease 314
characteristics and accomplishments 285
Christmas dinner with Robert Stephenson in Cairo 314
died one month before Robert Stephenson 315
joined by Robert Stephenson at *Great Eastern* launch 311
joins Robert Stephenson at Menai bridge site 308
much in common with Robert Stephenson 284–5
one of the greatest engineers 269
riding Cairo streets on a donkey 314
Brunton, William, inventor of *Mechanical Traveller* 90
Buckland, William, Professor of Geology
discusses coal formation with George Stephenson 290–1
Burstall
built road carriage 9
caught spying on *Rocket*'s progress 186
modifies *Perseverance* during supposed repairs 192
offered to build a locomotive for the L & M Railway 229
Scottish engineer 9
vital dates 9

cab drivers, pelt London carriage 52
caloric engine 71, 294–5
Camborne
Cornish mining centre 45
steam carriage 49–50
Trevithick's home 43
canals, haul heavy freight 126
Canning, George, Tory moderate 232
Cartagena meeting between Trevithick and Stephenson 150–2
cast iron
brittleness 33, 82
differences from wrought due to manufacturing process 327–9
excellent for casting 82
unsuitable for rails 82
wheels 33
Catch-me-who-can 58
admission charged to see 57
derailment last straw for Trevithick 58
exhibited in London 57
exhibition closed 58
first locomotive in the world 54
mere curiosity 58
only modest interest 58
Trevithick's last locomotive 59
Cerro de Pasco
Peruvian mining region 60
Trevithick restores prosperity to mines 132
Changing directions, detailed explanation 170-3
Chat Moss 242, 243, 255
cheek plate, of eccentric assembly 171, 172
Chester & Holyhead Railway
bridging Menai Strait major obstacle 304
Dee bridge disaster 306–7

Robert Stephenson appointed engineer 305
ultimately to link London with Ireland, via Holyhead ferry 305
Cheverton's gas-powered engine 27
Chittaprat, rebuilt as *Royal George* 97
civil unrest 130
clinker from burning coke 163
clinking, securing ends of boiler tubes 180
coal
burns more intensely than wood 71
mining an essential industry 104
prices affected by S & D Railway 4
Coalbrookdale
centre of iron industry 51
engineers amazed by Trevithick's new engine 51
first iron bridge 306
locomotive purportedly built 52
Trevithick visits 51
Cochrane, Lord Thomas
captured Spanish flagship 133
commanded South American navy 133
later Earl of Dundonald 133
returns to England 271
uses *Rocket* to test engine invention 274
warned of assassination plot by Trevithick 134
cockfights and dogfights 106
coke, burns less readily than coal 165
cold caulking 80
cold collation for opening-day guests 257
Colombian Mining Association
perceived opportunities for Forth Street Works 139
requested Robert Stephenson to visit Cornwall before departure 140
Robert Stephenson engineer-in-chief 135
Combination Acts, outlaw trade unions 111
convection currents 182
Cook's Kitchen mine 48
Cooper, Sir Astley, fears railways will 'destroy the noblesse' 276–7
corf-bitter 104
Corn Law
kept bread prices high 232
protest during opening day, L & M Railway 257
Cornwall
copper and tin mining 45
largest market for Boulton & Watt 42
resentment for Boulton & Watt 42
corrosion 326
corves 107
Costa Rica, Trevithick visits 134
cotton industry, Manchester at centre 4
Count Potocki, guest aboard Wellington's carriage 247
coupling rods 17, 149, 183
Coxlodge Colliery 94
cranked axle
description and function 66
early plan by George Stephenson 121
lack of manufacturing skills delay use 121
not original to Braithwaite & Ericsson 66
secret of *Novelty*'s smooth motion 66
used in *Planet* 275
Cree, John, killed in *Locomotion* explosion 2
Creevey, Thomas, account of train ride 234–5
Cropper, James, board member enemy of Stephenson 224
cross-head 47, 55
Cugnot, Nicholas, built first steam carriage 50
cupping, useless medical procedure 291

Cycloped 28
designed by Brandreth 26
gives second demonstration 77
horse falls through floor 77
not driven by steam 26
powered by horses 27
too slow to compete 29
cylinders
steam introduced at both ends 19
valves control steam 19
Wilkinson's boring machine 42

dalmatian dog, Turnbull's face speckled like 2
dandy cart 28
Darlington 124, 126
Darwin, Charles, similar views to Robert Stephenson on Professor Hope 137
Davy, Sir Humphrey, inventor of safety lamp, denies George Stephenson's claim 124
De Dunstanville, Lord and Lady 43
Dewley Burn 103
dies non, a non-day 191, 194, 200
Ding Dong mine 48
diving bell 72, 131
Dixon, John
appointed to Stockton & Darlington Railway 129
could not please Hackworth at Rainhill 82
disparaging remarks about *San Pareil*'s large appetite 214
disparaging remarks about *San Pareil*'s rolling motion 206
Hackworth 'must return to his preaching' 221
resident engineer, L & M Railway 82
scathing quote about Burstall 221
sinks into Chat Moss 242
Dodds, Ralph
bonus and promotion to George Stephenson 116
desperate to repair recalcitrant Newcomen engine 113
head viewer at Killingworth 113
improves power transmission with George Stephenson 120
patents improvements with George Stephenson 122
reaches agreement with George Stephenson over Newcomen engine 114
safety concern for repaired Newcomen engine 115
dog, of eccentric assembly 171, 172
Dolly Pit mine 107, 108
dome, to reduce priming 161–2
Drayton Park, home of Prime Minister Peel 290
driver, of eccentric assembly 171, 172
drivers
breaking company rules 32–3
dismissal for speeding 33
earned more than Hackworth 85
highly skilled 19
lubricating trains in motion 33
reversing directions during trials 169, 170
tampering with safety valves 34
usually escaped injury during explosions 36
Wakefield 20
drivers, by name
Cree, John 2
Gowland, William 205
Stephenson, James 34
Sunter, George 33
Duke of Northumberland 103
Dunstable, site of Telford's chalk-cutting 278
Earl of Abergavenny, salvaged by Braithwaite Sr 72
Earl of Dundonald (see Cochrane)
Earl of Strathmore, outraged by Davy 124
eccentric 19–20
eccentric assembly 170–3
edge rail 87
Ericsson, John 64
age at Rainhill 10

appearance 65
arrival in England 65, 71
borrowed money for trip to England 71
Braithwaite's junior partner 23
Bright's disease 297
caloric engine designed for large ship 294
congratulations and praise for *Monitor* 296–7
conversation with his physician 298
death 298
depth gauge among his achievements 292
designs screw-propelled ship for US navy 294
destruction of flame engine 71
did not hear of trials until later 23
easy-going 65
flame (caloric) engine 71
former Swedish army officer 23
leaves England for America 294
meets Braithwaite 72
military bearing 65
Monitor, designed for Unionists 295
needed source of income 72
New York prize for fire-engine design 294
portrait 64
reasons for choosing England 70
ship's propeller, Admiralty unimpressed 293
uncontrollable temper 65
used his army title 65
vital dates 11
withdraws *Novelty* from the competition 219
Esmeralda, Spanish flagship, captured by Cochrane 133–4
expansive operation of steam 167–8
explosions
Diligence, Edward Corner thrown 16 yards 34
firemen most vulnerable 36
flue tubes usual cause 36
Mechanical Traveller 90
rare before 19th century 37
rarely involved boiler barrels 36
Salamanca, driver was 'quite dead' 35–6
usual causes of death 36

Fairbairn, James, account of Robert's encounter with Trevithick in Cartagena 151
Fairbairn, William, tests tubular bridge models 304
fastest humans had ever travelled 254
'Father of locomotive' 15
feed pump 69
Fenton, Murray & Wood, locomotive builders 88
ferrule 179
fire engine of Braithwaite & Ericsson 74–5
fire engine, older name for atmospheric engine 37
fireman's responsibilities 18, 169, 208
firemen, Edward Corner, John Gillespie 34
fires, draughts and chimneys 160
fishbelly rails 122
Fitzroy Square 59, 60
flame engine, see caloric engine
flue tube
catastrophic failure 36
function 13
Novelty's 70
poor heat circulation in surrounding water 215
straight-through and return types 13
uneven heating 36
flywheel, needed for single-cylinder engines 45
Forth Street Works
barely any locomotive advances during Robert's absence 148
continue development with new *Planet* class 275
foundation 134
George Phipps 177
post-Rainhill locomotives 233

Forth Street Works – *cont*
Robert reminded of plight while in South America 148
William Hutchinson 177
fusible plug 212

Gateshead, second locomotive in the world 57
Gauge Commission and Gauge Act 288–9
Geordie lamp 124
Gerard, J. M.
dies in poverty in Paris 272
in Cartagena with Robert Stephenson 150–1
returns to England with a new mining scheme 271
Gilbert, Davies
carriage experiment with Trevithick 48
informed of Trevithick's Welsh prospect 53
Member of Parliament, helps Trevithick with his appeal 272
Trevithick's friend and confidant 48
Gillespie, John, died in *Diligence* explosion 34–5
Globe, passenger locomotive for S & D Railway 300
Gooch, Daniel, brother of Thomas, locomotive designer for GWR 288
Gooch, Thomas
assists Robert on the London & Birmingham Railway 275
guest aboard *Titania* Norway trip 314
government, imposes heavy taxes 111
Gowland, William, probably crewed *Sans Pareil* 205
Grand Allies
impressed by George Stephenson 117
invest in Newcomen engine 113
owned, or controlled, many northern collieries 94
pay George Stephenson a retainer 125
Grand Junction Canal 276
Grand Junction Railway, dispute between Locke and Stephenson 279–80
Great Western Railway 285, 287, 288

Hackworth 83
boiler plans for *Sans Pareil* discussed by Stephenson 183
born same village as George Stephenson 12
brief mention of Cree's death 35
builds *Royal George* 97
builds *Sans Pareil No. 2* 302
built *Sans Pareil* himself 82
cold-caulks leaking boiler seams 80
considered for job with S & D Railway 84
contented life 96
death 304
demanding job 83
designs *Globe* for passengers service 299
devout Methodist 84
difficulties in building *Sans Pareil* 83
discovers leaks in *Sans Pareil*'s boiler 26
disputes accuracy of scales with judges 204
dour looking 84
enters into partnership to build locomotives 301
excessive use of blast pipe 207
father's death 86
favoured return-flue boilers 13
finally severs ties with S & D Railway 301
first superintendent, S & D Railway 85
five years junior to George Stephenson 12
formally requests a second trial 219
good life at Wylam 94

helped build *Puffing Billy* 92
helps build Wylam locomotives 91
ill humour at Rainhill 82
implications of winning at Rainhill 210
improves spring-loaded safety valve 98
joins S & D Railway 96
large contract for London & Brighton Railway 301
leaves Walbottle Colliery 96
leaves Wylam Colliery 95
marriage 91
modest salary from S & D Railway 85
new relationship with S & D Railway as contractor 300
no time to test *Sans Pareil* before trials 83
not ambitious 96
promotion to foreman blacksmith at Wylam 86
punctilious letter to Pease 84
rather dour 26
refused to work Sundays 85, 95
requests judges for more time 190
resolves rope-haulage problem for S & D Railway 97
Samson, built for Canadian colliery 301, 344
seven-year apprenticeship at Wylam 86
small inheritance used up on *Sans Pareil* 83
stayed at school until 14 85
talented blacksmith, metalworker, boilermaker 85
talents recognised by father 85
testimonial from Walbottle Colliery 84
tries business 96
views on the Sabbath well known to Wylam employer 94
vital dates 12
worked briefly for George Stephenson 96
worked with Hedley 88
works on leaking boiler 77
writes to directors of L & M Railway after Rainhill 228
Hackworth & Downing, see Soho Works
Hackworth, John, son of Timothy
delivers locomotive to Russia 301
letter challenging Robert Stephenson to locomotive trial 303
Hall, Bruce Napier, officer who saved Trevithick's life 151
Harveys of Hayle 43
Hawthorn, Robert, challenger to Stephenson in feats of strength 105
Hedley, William
supervises second Wylam locomotive 91
test carriage traction experiments 88
viewer at Wylam 87
Henderson, Francis (Fanny), George Stephenson's first wife 109
Heppel, Kit, talks with Stephenson about Newcomen engine 113
Herapath's Railway Journal publishes John Hackworth's challenge 303
Hetton Colliery, George Stephenson builds railway 125, 126
High Pit mine 113
Hindmarsh, Elizabeth, George Stephenson's second wife 130
Homfray, Samuel, owned Pen-y-darren engine 54, 57
Hope, Professor Thomas, taught Stephenson at university 137
horse-drawn wagons, convey S & D Railway passengers 3
horses
cheaper alternative sought 86
demise predicted 56
haul ten times more on rails than roads 126
haulage capacity for rail wagons 29

horses – *cont*
haulage compared with locomotives 29
integral part of everyday life 28
jobs of handlers threatened by locomotives 92
used on wagonways 86
Hudson, George, 'Railway King' 267-9
Huskisson, William
death 261
delivers speech to his constituents in Liverpool 244
reconciliation with Wellington 250–1
reform proponent 233
relegated to back benches 233
revises will 256
seriously wounded by *Rocket* 252
Tory moderate, Member for Liverpool 232
Hutchinson, William, manager at Forth Street Works 177
Huyton village 240
hydraulic permit 184

inclines
for hauling wagons 3
on S & D Railway 3
Whiston and Sutton, flanking Rainhill track 217
income tax 111
industrial accidents 116
industrial espionage 186
industrialization, Britain led way 4
Institution of Civil Engineers
Ericsson sent flame-engine manuscript 71
Robert Stephenson President 315
Robert Stephenson joins 275

James, William, lawyer, land agent, railway advocate 127, 135
Jameson, Robert, taught Stephenson at university 136
Jeaffreson, Cordy J., Robert Stephenson's biographer 155
Jenyns, the Reverend Leonard, account of winter 231
Jessop, Josias, consultant engineer above George Stephenson 146, 147
Jones, Catesby, *Merrimac*'s commander 296
judges
allow time for repair of *Novelty*'s bellows 76
appointed by directors 10
decide appropriate test load 18
decide regulations for trials 30
determine *Sans Pareil* is overweight 204
grant Hackworth extra time 26
grant Hackworth more time 77
grant Hackworth more time and plan 4-day pause 190
Liverpool meeting on *dies non* 195
reasons for declining second trial for *Sans Pareil* 219–20
report of *Sans Pareil*'s breakdown 213
report of *Sans Pareil*'s performance 214
report on *Novelty*'s third breakdown differs from Stephenson's 219
rule *Sans Pareil* ineligible, but decide to let her compete 204
visit *Sans Pareil* 77
were expected to give *Sans Pareil* a second trial 214

Kemble, Francis (Fanny) 237
account of locomotive journey 236, 238, 241–2
guest of railway for opening ceremonies 244
sets off on walk with George Stephenson 240
stage debut as Juliet 158
thoughts on George Stephenson 243
vital dates 236
Kennedy, John, Rainhill judge 10

Killingworth Colliery
Blenkinsop once viewer 87
employed George Stephenson 94
Pease visits to see locomotives 128
Rocket's successful test run 187–8
test site for *Rocket* 18
William James visits 127
Kilsby tunnel, formidable obstacle 282
King George IV 247
Kitching, William, believed Hackworth unfairly treated 227

Lancashire Witch 166
limited steaming capacity 174
pistons worked expansively 167
resemblance to *Rocket* 167
Robert Stephenson took leading role 165
Rocket's antecedent 164
Lardner, Dr Dionysius, author, lecturer 14
laudanum 256
leeches, for medicinal use 291
Liddell, Sir Thomas
authorises locomotive construction 118
confident in George Stephenson 120
encourages George Stephenson 117
later Lord Ravensworth, one of the Grand Allies 94
plans to build a locomotive 94
Lima, Trevithick's arrival 95
Lincoln, Abraham, receives proposal from Ericsson 295
Liverpool 4, 244
Liverpool & Manchester Railway
Act of Parliament 19
announces Rainhill trials 7
armed opposition to survey 144
Bill defeated 145
Brandreth former director 26
business growth 265
committed to steam power 4
compensates widow 7
decline Burstall's offer to build a locomotive 230
directors arrive for trials 8
dispenses with James's services 135
divided between locomotive & stationary engines 4
early profits outstanding 264
hire Jessop as consulting engineer 146
hire the Rennies above George Stephenson 102, 145
Jessop monitors George Stephenson's work 147
John Dixon resident engineer 82
journey along the line 238, 240-2
Millfield Yard 9
no longer needed stationary engines 218
offers £500 prize 7
opening day 245-61
orders for more engines like *Rocket* 224
primarily intended for freight 262
reinstate George Stephenson after second Bill is passed 147
smoke stipulation 19
statistics for first month in operation 263
stipulate two safety valves 37
stipulations for competition 7
stipulations for contestants 31, 320–1
total cost of project 244
two *Novelty*-type engines ordered against Stephenson's advice 224
two-way traffic 6
Vignoles resigns after difficulties with George Stephenson 147
vote of thanks to George Stephenson 235
Liverpool and Manchester Canal 144
Liverpool Chronicle, enthusiastic account of *Novelty* 25
Llangollen, site of Rainhill re-enactment in 2002: 163, 193
local militia, to maintain national law and order 112

Locke, Joseph
  became President, Institution of Civil Engineers 249
  becomes *persona non grata* with George Stephenson 280
  driving *Rocket*, cannot avoid collision with Duke's carriage 252
  named one of the three great engineers 284
  reconciled with Robert Stephenson 307
  replaces Vignoles at L & M Railway 147
  testifies for Robert Stephenson at Dee bridge inquest 307
  that George Stephenson saw *Novelty* as only serious rival 75
  works on Grand Junction Railway 279–80
Locomotion
  hauls inaugural train 2
  replica's boiler modified 36
  tyre failure attributed to coupling rods 183
locomotives
  Blenkinsop's is first commercial success 90
  chuffing sound 21
  demand attention 163
  drivers highly skilled 19
  earliest unreliable 2
  early ones used coal 18
  Edward Pease realises potential 128
  evolution not linear 89
  explosions 2
  fireman's responsibilities 18, 169, 208
  first in world 54
  frightened livestock 2
  haulage capacity compared with horses 29
  how slowed down 20
  improvements after Rainhill 233, 275
  information not freely shared 89
  lubrication in motion 33
  minor developments during early 1800s 149
  modern ones have variable cut-off 168
  poor steaming capacity of early ones 34
  second in world built at Gateshead 57
  slow acceleration 21
  sparks and smoke from chimney 2
  Trevithick builds first in world 54
  two operators required 18
  without brakes 19
  Wylam's activities 91
locomotives, by name
  Agenoria 5
  Arrow 233, 235, 246, 259
  Blücher 122
  Catch-me-who-can 57, 58, 59
  Chittaprat 97
  Comet 233, 246
  Cycloped (horse powered) 9, 26, 27, 28, 29, 77
  Dart 233, 235, 246
  Diligence 34, 35
  Globe 300
  Hope 97, 148
  Ixion 289
  Lancashire Witch (*Liverpool Travelling Engine*) 164-5, 166–8, 174, 175
  Locomotion 2, 35, 36, 97,148, 183
  Mechanical Traveller 90
  Meteor 233, 246
  My Lord 118–20, 122
  North Star 233, 246, 249, 255, 259
  North Star (broad gauge) 288
  Northumbrian 234, 236, 240, 241, 246, 248, 249,250, 252, 254, 255, 258, 262, 275
  Novelty 9,10, 11–2, 18, 23, 24–6, 27, 29, 30, 61, 65–70, 73, 74, 75–76, 78, 122, 190, 191, 192–4, 195–7, 198–199, 202, 205, 206, 209, 213, 215, 216–7, 218–19, 221, 222, 224, 227, 294
  Perseverance 9
  Phoenix 233

Planet 275
Puffing Billy 92–3
Queen Adelaide 224
Rocket 9, 10, 14, 15, 17, 18, 19, 20–2, 24, 25, 68, 69, 70, 75, 76, 77, 82, 83, 92, 100, 102, 134, 158, 159, 160, 161, 162, 163–4, 167, 168–74, 176–90, 192, 193, 195, 196, 197–8, 202, 204, 205, 206, 207, 208, 209, 210, 211, 213, 214, 215, 217, 218, 221, 222, 223, 224, 229, 230, 233, 234, 246, 249, 251, 252, 259, 274, 275
Royal George 97–98
Salamanca 35
Samson (colliery engine) 220
Samson (main-line engine) 224
Sans Pareil 9, 12, 13, 14, 17, 24, 26, 29, 68, 69, 70, 75, 77, 79, 80, 82, 190, 191, 192, 193, 200, 201–15
Sans Pareil No. 2 302
William IV 224
London
enormous impact of railways 283
largest coal market 94
locomotive exhibited by Trevithick 57
locomotive's potential initially not realized 58
no rail tracks at time of Rainhill 18
rapid growth 94
London and Birmingham Railway
Bill finally passed 278
fierce opposition from canal and landowners 276
formidable challenges and statistics 281
Kilsby tunnel's staggering statistics 282
major project, linking capital with industrial north 275
monopoly claims force boycott of Forth Street Works 284
London Fire Brigade 75
London steam carriage 52–53
Longridge, Michael
becomes partner in Forth Street Works 134
burdened by Forth Street Works, hopes for Robert's return 148
concerns for Robert's South American trip included in a letter 141
high regard for Robert Stephenson's abilities 141
ironworks produces wrought iron rails 130
partner of George Stephenson 82
Lord Castlereagh, Minister of War 112, 232
Lord Derby, opposition to L & M Railway 144
Lord Sefton, opposition to L & M Railway 144, 234
Lord Wharncliffe, Stuart Wortley
one of the Grand Allies 94
reassures Robert Stephenson over railway Bill 278
Wellington's guest aboard carriage 247
Losh, William, worked with Stephenson on rail improvements 123, 130
lubricating trains in motion 33

MacAdam, road improvements 126
Maidenhead bridge 285
Manchester 4
Manchester Guardian, dispute with *Mechanics' Magazine* 225–6
manometer 162
Manufactured products, purchased by working class 73
Markoe, Dr Thomas, Ericsson's physician 298
Marquis and Marchioness of Salisbury 247
Mechanics' Magazine
critical of Rainhill organisers over

workshop 197
critical of *Rocket* on trial 22
details of *Novelty*'s boiler 68
dredges up the past to attack George Stephenson 223
faint praise for Stephenson, predict *Rocket*'s eclipse 223
no description of *Cycloped* 27
responds to *Manchester Guardian's* attack 226
understood judges would give *Sans Pareil* a second trial 214
Vignoles a correspondent 78
medicine, in primitive state 255–6
Menai Strait 304
mercury, used to extract silver 132
mercury gauge 162
Merrimac, Confederates ironclad warship 295
Mersey and Irwell Navigation Company
canal operators, opposed to L & M Railway 144
Messrs Rastrick and Wood
manned their posts during *Rocket*'s first 35-mile journey 189
take their positions prior to start of *Rocket*'s trial 164
timekeepers at Rainhill 168
methane, cause of mine explosions 123
Middleton Colliery, Blenkinsop's railway 88
Millfield Yard 10
mine captain 42
miner's safety lamp 123
Monitor, ironclad warship designed by Ericsson 295–6
Montrose, Scotland 111
Multi-tubular boiler 175–6
Murdock, William, employed by Boulton & Watt 45, 47
Murray, Matthew, designed Blenkinsop's locomotive 88

Napoleon, final defeat 73
Napoleonic Wars
began 1793: 53
bread prices doubled 104
causes inflation 111
civil unrest follows peace 130
financial drain on Britain 111
income taxes introduced 111
inflate price of horse fodder 86
National Railway Museum
Agenoria displayed 5
restored *Novelty* replica for Rainhill re-enactment 193
Natural History Museum, prehistoric reptiles 93
navvy 282, 349
Nelson, Ned, fights George Stephenson 108
Newcomen engine 38
became larger with time 40
brass cylinders in early ones 40
first 40
heavy on fuel 40
installed at High Pit mine, near Killingworth 113
low efficiency 41
numbers built 40
smoothing bores manually 40
Newcomen, Thomas, developed engine 39–40
newspaper claims rules changed 159
Newton town and viaduct 241
northern mills 73
Northumbrian
comparisons with *Rocket* 234
hauled Wellington's train on opening day 246
ridden by Fanny Kemble 236
speeds to Manchester with Huskisson 254
Novelty 11
air-compressing system 11, 66
air pump 67
air pump had no contemporary accounts 192
air pump mystery unravelled 193–4

allowed time for bellow repairs 76
appearance 11
bellows failure during first trial 76
boiler details 67–70
boiler failure ends third trial 219
burst pipe ends second trial 197
captures the crowd 26, 216
carried its own fuel and water 30
copper steam chamber 11
cylinder bore 69
dependent on forced air supply 67
fast start to first trial 75
faster than *Rocket* on first day 25
first locomotive with cranked axle 66
first trial begins 75
flying start to second trial 196
forced air supply increased with speed 67
generation of steam 69
implications of Rainhill victory 76
last-minute checks 65
lightest competitor 12
men working on boiler before 3rd trial begins 217
no railway tracks for testing in London 18
re-tested after Rainhill, without success 224
reaches 28 mph first day 24
replica with pulsed air pump 193
replica's performance inferior to original's 194
second trial begins 196
slow start to third trial 218
smooth action 25
resemblance to Braithwaite & Ericsson's fire engine 74
third trial begins 217
thrills spectators and passengers after pipe repair 198–9
untested before trials 14
weight 70

Olive Mount 239
crowded with spectators 249
described by Fanny Kemble 238
workman killed at railway cutting 7

Opening day, L & M Railway 248
cavalcade stops at Parkside and some men stretch their legs 250, 251
cold collation for guests 257
crowds converge, hotels full 245
estimates of crowd 259
honourable guests 247
Huskisson beside Duke's carriage as *Rocket* bears down 251
Huskisson struck and badly injured 252–53
law enforcers, in anticipation of trouble from radicals 246
political protests 257
press coverage 249–50
Prince Esterhazy scrambles to safety 251
some drivers as distinguished as many of the guests 248
stranded trains hitched together 259
trains flags correspond to ticket colours 246

Order of St Olaf, Grand Cross awarded to Robert Stephenson 315

Parkside, where locomotives took on water 250, 251
Parliament, smoke legislation 19
pauperism 105
pay-day at the colliery 105–6
Pease, Edward
appraised of locomotives 126
becomes partner in Forth Street Works 134
down-to-earth 125
first meeting with George Stephenson 124
founding director of S & D Railway 126
grooms George Stephenson 129
impressed by Hackworth's morals 85
interviewed Hackworth for job 84

Pease, Edward – *cont*

invites George Stephenson to survey S & D Railway 127
preferred rail link than canal between Stockton & Darlington 126
recollections of meeting George Stephenson 125
reminiscences of difficult times of S & D 150
shrewd businessman 125
strait-laced Quaker 85
views on rail roads 126
wealthy and influential businessman 83
woollen trade interests 125
Pease, Joseph, Edward's son 144
Peel, Sir Robert
concurs with Wellington in cancelling ceremonies 254
guest aboard Wellington's carriage 247
weekend guests included Stephenson and Buckland 290
Pen-y-darren locomotive 55
appearance 54–5
built for Samuel Homfray 54
first in world 54
frequently broke rails 56
had smooth wheels 88
plate rails used 87
test-run successful 54
undistinguished career 56
perpetual motion
delusion based on misunderstanding 110
among Rainhill proposals 8
description 109
intrigues young George Stephenson 109
Perseverance
appearance 221
carried its own fuel & water 30
damaged in transit 10
modified by Burstall during supposed repairs 192
pedestrian performance 221
temporarily out of commission 26
Peru, under Spanish control 132
Phipps, George
draughtsman at Forth Street Works 177
guest aboard *Titania* on Norway trip 314
recollections of fitting *Rocket*'s boiler tubes 179
recollects building construction of *Rocket* 177
piston rod 47
pistons 19
Pitt, William, government imposes heavy taxes 111
plate rails 87
political protests 257
press controversy after Rainhill 222–3, 225–6
pressure 162
priming 161–2
Prince Esterhazy, guest aboard Wellington's carriage 247

Queen Victoria 283

rails
cast and wrought iron 33
edge 87
fishbelly 122
how the 4ft 8 ½ in. gauge originated 288
initially only cast iron 122
iron 87
plate 87
standard and broad gauge 287–8
variety of shapes and sizes 87
railway mania 267
railway surveying 128
railways
enormous impact on London 283
explosive growth 262
heightened expectations 267
impact summarised by Robert

Stephenson 315–6
mileage and market statistics 266
stagecoach's demise 265
the bubble bursts 269
Rainhill, northern hamlet 1
Rainhill trials 23
announcement 7
assigned loads 30
atrocious weather on day two 77
coke stipulated 19
company stipulations 31, 320–1
enterprising landlady 23
expected to last three days 9
fanciful proposals received 8
few independent accounts 78
judges' plan four-day pause 190
judges regulations 30–1
newspaper claim rules changed to favour Stephensons 159
no activity on the Sabbath 200
numerous enquiries 7

October, 6, first day 1
October, 7, second day 62
October, 8, third day 157
October, 9, a *dies non*, a non-day 191
October, 10, fourth day 195
October, 12, a second *dies non* 200
October, 13, fifth day 203
October, 14, sixth and last day 216
only five competitors 8
perpetual motion proposed 8
press controversy in aftermath 222–3, 225–6
re-enactment at Llangollen in 2002: 163, 193
refreshments inadequate 22
rules and conditions distributed on cards 158
short notice given 7
single track only used 8
special constables 8
Stephenson had advanced notice 7
stipulations drafted 7
three judges appointed 10
why necessary 2
Rastrick, John Urpeth, respected engineer
appointed judge 10
built *Agenoria* 5
consulted on fixed and locomotive engines 5
sketch of *San Pareil's* boiler 203
Redruth, Cornish mining centre 45
reform movement 130–1
Reform Bill 272
regulator valve 20–1, 170
Rennie, George & John
hired above George Stephenson by L & M Railway 102, 145
refuse to work with George Stephenson 146
return-flue boilers
advantage 13
limited steaming capacity 220
more difficult to build 13
reversing directions 20, 170–3
Richardson, Thomas
assists Pease to groom George Stephenson 129
becomes partner in Forth Street Works 134
London banker 128
Pease's cousin 128
seeded idea of South America in Robert Stephenson 139
rioting, against high food prices 112
riveting 80–1
Robert Stephenson and Company, see Forth Street Works
Roby embankment 240
Rocket 17
alarming at speed 198
almost lost at sea 189
antecedents 164
blast pipe experiments after test run 188–9
boiler plates of 'Best RB' 82
boy sticks to wet barrel 21
completes trial successfully 190
Rocket – cont

copper water jacket for firebox 17, 182
coupling rods rejected to avoid power loss 183
crew lost time on turnarounds 211
cylinder bore 69
demonstrates speed without tender 197
designed for winning 168
driver on opening of trials 20
driver's judgement changing directions 173
early arrival at Rainhill 10
entered by Booth and the Stephensons 100
entertains crowd after *Novelty*'s first breakdown 76
factory built 18
finished weeks before trial 18
firebox 17, 181–2
first leg of trial slow 169
funding 15
heat-exchange area compared with *Sans Pareil*'s 215
more stays added to boiler 186
no coupling rods 17
Phipps's recollections of construction 177
pistons oblique 17
predicted eclipse by its Rainhill competitors 223
preparation by crew before start of trial 164
pressure testing halted 185–6
problem fitting boiler tubes 178–9
races up Whiston incline 217
reaches 30 mph 77
reaches 32 mph 198
remains displayed at Science Museum 92
reversing mechanism, attention in designing 169
rolling motion 22
runs over Huskisson 252
smokes first day of trials 22
speeds during trial 190
start of trial 164
Stephenson could calculate specifications 180
test-run at Killingworth reported to Booth 187–8
time taken to raise steam 163
two-wheel drive 17
used to haul ballast at Chat Moss 230
weighed and assigned load before trial 159
weight distribution 181
rope haulage, problems for Hackworth 96
rotten boroughs 234, 345
Royal George 98
built by Hackworth 97
cylinder bore 203
description 97–8
direct power transmission 97
'finest in world' 99
John Stephenson impressed 99
most powerful locomotive 99
pre-heated boiler water 98
spring-loaded safety valve 98
Rugby, near Kilsby 281
SS *Ericsson*, built for caloric engine 294
SS *Ogden*, built for Ericsson's propeller trials 293
SS *Stockton*, second vessel for Ericsson's propeller trials 293
Safety valve
tampering with 34
to vent excess steam pressure 34
details 161
early use by Trevithick 53
out of driver's reach 37
spring-loaded 98
Salamanca, 35–36
San Martin, sank in Peru 131
Sanders, Joseph
campaigns for rail link between Liverpool & Manchester 127
converses with Huskisson during opening ceremonies 250

influential Liverpool merchant 127
invites James to survey L & M Railway 127
Sankey viaduct and canal 240
Sans Pareil 13
assigned load 204
based upon *Royal George* 202
boiler casing caps top of flue tube extension 202
boiler leaks fixed and test run successful 201
calamity half-way through trial 212
collides with Winan's vehicle 29
coupling rods 14
cracked cylinder casting discovered after trials 227
cylinder bore 69, 203
demonstrated to L & M Railway after repairs 228
driver and fireman at opposite ends 205
effectively without springs 221
efficient reversing mechanism 211
extensive leaks in boiler 80
extra long grate difficult to stoke 208
fastest start of all contestants 205–6
feed pump concerns crew 211
fireman has to ask driver to activate feed pump 209
fireman's difficult and hazardous job 207–8
flue tube extension increased furnace area 202
forceful blast pipe wasted fuel 207
four-wheel drive 14
fusible plug 'melted out near Grand Stand' 213
gave long service 229
Hackworth's boiler plans discussed between Stephenson and Booth 183
heat-exchange area compared with *Rocket*'s 215
hot boiler at start of trial 204
leaking boiler delays start 26
likely crew for trial 205
many shortcomings made it unsuitable for main-line service 220
massive appearance 12
not tested before Rainhill 14
overweight 204
purchased by L & M Railway 229
pushed tender from front 12
Rastrick's sketch of boiler 203
return-flue boiler 12
rolled 'like an empty beer butt' 206
try-cocks difficult to reach 209
was soundly outperforming *Rocket* during her trial 211
Santa Anna, Colombian mining region 142
school leaving age 85
Science Museum, displays
Puffing Billy, oldest locomotive in world 92
Rocket's remains 92
Trevithick engine 47
Scientific American predicts failure of SS *Ericsson* 294
Séguin, Marc, patented multi-tubular boiler before Rainhill 176
separate condenser 41
Shildon, S & D Railway's workshop at 83
shipping, only way to move heavy goods 10
sight gauge 208
size constraint of flue-tube boilers 220
Smiles, Samuel, Stephenson's biographer
compares Stephenson & Watt 15
erroneous account of fitting *Rocket*'s tubes 178
perpetuates Stephenson myth 15
visits Robert Stephenson 311
smokebox 233–4
Soho Works 301
South Seas whaler 95
St Etienne Railway 175
Staffordshire potteries 73

stagecoaches
fares compared with railways 263
George Stephenson's account 265–6
romance and reality 265–6
standard gauge 287
Stannard, R., recollections of *Rocket* 21
stationary engines for hauling wagons
greater reliability over early locomotives 3
L & M Railway divided between them and locomotives 4
no longer needed on L & M Railway 218
steam
cause of death during explosions 36
high and low pressure 37
operating expansively 167–8
steam engine
distinct from atmospheric engine 37
high-pressure developed by Trevithick 43
portable, designed by Trevithick 47
working principle 37
steam gauge, early use by Trevithick 53
steam locomotive, first built by Trevithick 15
steam springs 123
steel springs, unavailability 123
Stephenson, George 101
acceptance speech for safety lamp award 124
advanced notice of Rainhill trials 7
advocate of locomotives 5
ambitious 106
appointed engineer S & D Railway 130
appointed surveyor S & D Railway 127
astute observer 106
attacked in *Mechanics' Magazine* 229
becomes brakesman at Dolly Pit mine 107
birth of son 110
bonus and promotion for Newcomen engine repair 116
buys donkey for Robert to ride to school 117
celebrates with Robert at Dun Cow 282
childhood 103
claims of having railway monopoly 284
clock repairs 110
competitive 105
completes second locomotive 122
considers emigrating to America 112
corf-bitter 104
courts Fanny Henderson 108
critiques consultant's report 6
dealings with George Hudson 268
death and funeral 292
death of second wife 291
declined a knighthood several times 290
defeat of Railway Bill 102, 145
did not go drinking like most of the men 105
discusses coal formation with Buckland 290–1
dismantles engine for pleasure 106
does not delegate 130
drives *Locomotion* 2
drives *Northumbrian* with wounded Huskisson 254
drove *Rocket* first day of trials? 20
early life 103, 104
early locomotive 121
early visits to Wylam to see locomotives 94
enemies on the board include James Cropper 224
engineer-in-chief, L&M Railway 5
enjoyed Liddel's confidence 120
erects atmospheric engines 116
evening classes 106–7
falls out with Locke over Grand Junction Railway 280
familiarity with atmospheric engine 106
father's injury 111

fiercely proud 5
fights Ned Nelson 108
financial security 125
first locomotive 118
first marriage 109
first serious illness 291
first wife dies 110
Forth Street Works partnership 134
fortunes begin to change for better 112
gave dandy cart idea to Brandreth 28
gruelling cross-examination 145
helps fund *Rocket* 15
Hetton Colliery Railway built 125
hostility to Vignoles 102
impresses Grand Allies 117
improvements in power transmission 121
incensed by consultant's report 5
initially sees *Novelty* as only serious rival to *Rocket* 75
intrigued by perpetual motion 109
invents safety lamp 123
invited to meals while surveying railway 129
invited to superintend Boulton & Watt engine in Scotland 110
irrational jealousy 15
lacked qualifications 102
learns about Watt's engine improvements 109
learns school lessons with Robert 117
left Dolly Pit mine 109
London engineers disdained 102
Losh relationship ends 130
low point in life 111
made enemies of rivals 15
makes suit for Robert 117
marries first wife 109
marries second wife 130
marries third wife 292
mechanical abilities outstanding 106
meeting with Pease 125
military service avoided 112
Montrose job 111
moves to Killingworth 110
much in demand after L & M Railway success 274
natural curiosity 106
no schooling 103
northern legend 100
not mercenary 5
not meticulous 279
Novelty has 'no goots' 76
Novelty-type engines ordered against his advice 224
offers services to S & D Railway 125
opposes fixed engines for L & M Railway 5
part ownership of colliery 125
patents improvements with Dodds 122
pays father's debts 111
perseverance 107
practises lessons 107
premature wear on rope eliminated 112
presented with award for safety lamp 124
prominent figure at Rainhill 100
promotion to brakesman 107
promotion to engineman, Walbottle Colliery 105
promotion to engine-wright 116
promotion to fireman 105
quote 'made man for life' 105
realises limitations imposed by rails 122
recommends wrought iron rails to S & D Railway 130
recommends wrought-iron rails 6, 33
resolves Newcomen engine problem 115
retirement at Tapton House 289
saves for Robert's education 117
saw things empirically 138
scarecrow built 116

Stephenson, George – *cont*
second child born 110
second wife a former sweetheart 130
self-acting incline installed 117
sets off on walk with Fanny Kemble 240
shoemaking and repairs 107
simple traction experiment 119
soldiering 130
sound knowledge of engines 114
stagecoach journey recounted 265–6
steam springs 123
S & D Railway opening 145
strong accent derided 102, 145
sundial built with Robert 118
surveys S & D Railway with son and Dixon 129
takes interest in Newcomen engine installation 113
takes job in Scotland 111
taught his son practical engineering 137
travels 20,000 miles by road in busy two-year period 280
turned Fanny Kemble's head 243
unenthusiastic about Robert's South American plans 140
uses edge rails 119
visited Coxlodge Colliery to see Blenkinsop locomotives 94
vital dates 5
vows never to change 124
Walker's estimates challenged 6
walks to Scotland 111
wealth and public recognition mellow him 281
withdraws from Grand Junction Railway 280
worked at Killingworth Colliery 94
working on second locomotive 120
works on recalcitrant Newcomen engine 114
works with Losh on rail improvements 123
writing a challenge 6
Stephenson, James, elder brother of George 34
Stephenson, Robert 16
account of *Novelty*'s third failure differs from judges' 219
appearance 15
appointed engineer, Chester & Holyhead Railway 305
appointed engineer, London & Birmingham Railway 275
assists father to survey S & D Railway 129
attends village school 117
attention to *Rocket*'s reversing mechanism during construction 169
attributes all his achievements to his father 316
bamboo cottage in Colombia 138–139
becomes partner in Forth Street Works 134
blast pipe experiments on *Rocket* after test run 188–9
built *Rocket* 15
buried at Westminster Abbey 315
calculating *Rocket*'s specifications 180
chalk-cutting challenged 277–8
chance encounter with Trevithick in Cartagena 151
'charmed all' 17
childless 310
Christmas dinner with Brunel in Cairo 314
Colombian venture futile 142
conceptual thinker, analysed engineering problems 138
coolness to Trevithick unexplained 152
could command workmen like an army general 283
declined knighthood 311
Dee bridge disaster 306–7
departure for South America 140
designed *Rocket*'s firebox 181
designs a box-girder bridge to span

Menai Strait 304
desk with secret drawer 312
developing engines but interested in building railways 275
dies, from 'dropsy' 315
discusses Hackworth's boiler plans with Booth 183
dogged determination 15
drives *Phoenix* on opening day 248
Dun Cow celebrations after Kilsby tunnel 282
Edinburgh University for one term 136
ends friendship with Locke out of loyalty to father 280
escapes aboard *Titania* 311
estimates *Rocket*'s weight 188
falls into Thames mud 311
formidable proponent of railway Bill 277
gives Trevithick fare home from Colombia 153
good terms with father 136
Hackworth seen as serious contender 183
health failing 312
home from Norway, makes a brief recovery 315
honoured in Norway 314–5
hounded by railway speculators 311
influenced by Trevithick about locomotives while in Colombia? 155
joined by Brunel at Menai bridge site 308
joins Brunel at launch of *Great Eastern* 311
Kilsby tunnel's enormous challenge 282
last letter home from South America 150
leading role in *Lancashire Witch* 165
letters to Booth about progress on *Rocket* 179
loses strong accent 136
married just before trials 15
Member of Parliament for Whitby 311
modesty 15
moves to London with his wife 279
never remarried 310
new thoughts on locomotives on return to England 155
night drive to Dunstable to check chalk-cutting 278
occupied by London & Birmingham Railway 281
plagued with self-doubt 283
President, Institution of Civil Engineers 315
problems with unruly Cornish miners in Colombia 143
puts brave face on his loneliness 310
reasons for departure to South America 135–6
reminded of plight of Forth Street during absence 148
responsible for *Rocket*'s construction 177
Rocket's test run at Killingworth reported to Booth 187–8
sails for Egypt aboard *Titania* 313
seeks father's approval for South American venture 140
sentimental journey to birthplace 312
served apprenticeship with Nicholas Wood 137
shipwrecked in hurricane after witnessing cannibalism 154
sought advice of others 15
South America's irresistible appeal 140
stays in South America 3 years 142
sundial seen, years after 312
taken gravely ill in Norway 315
takes Cornish miners to Santa Anna 142
tours Britannia bridge with friends on *Titania* trip 313
Stephenson, Robert – *cont*

unrecognised by old friend 312
vital dates 15
warned about Bedlington Iron Works 82
weakness of chest 140
wife's illness and death 310
Stephenson, Robert, George's father 103
Stimers, Alban C., *Monitor*'s chief engineer 296
stock exchange
purported fraud by Sir Thomas Cochrane 133
railway capital 266
Stockton & Darlington Railway
affects coal prices 4
appoints Dixon to assist Stephenson with survey 129
capital and running cost estimates 5
cavalier drivers 32
choose locomotives 127
clamour for shares 4
coal haulage by horses 29
cost estimates for fixed & locomotive engines 5
dismiss drivers for speeding 33
downturn in fortunes 150
drivers paid on commission 32
Edward Pease influential board member 83
essentially one-way traffic 6
experiences provide stipulations for trials 31
first public railway 2
fortunes improve after first difficult years 300
George Stephenson offers his services 125
Hackworth considered for job 84
Hackworth joins 96
Hackworth resigns 301
Hackworth becomes a contractor 300
horse-drawn wagons for passengers 3
inaugural banquet 4
initial success 4
large sum given by Pease 84
limited workshop facilities at Shildon 83
Locomotion first engine in service 97
locomotives and horses 3
mixed traffic 2, 3
passengers only conveyed by horses 29
Pease founding director 126
primarily for coal 3
rope-haulage problems 96
shares soar after opening 4
shortages and breakdown of locomotives during early days 148
single track 3
speed limits imposed 33
spurs railway speculation 4
thirty-one miles long 3
traversed by two steep hills 6
triumphant opening 2, 145
visited by Walker and Rastrick 5
wheel breakages high 33
Sunter, George, cavalier driver 32–3
surface area and size relationship 175–6
Swedish Railway Museum
loans *Novelty* replica for Rainhill re-enactment at Llangollen 193

Tapton House 289
Teague, Captain, meets Uville aboard ship 61
Telford, Thomas, eminent engineer
built a bridge across the Menai Strait 304
used similar chalk-cutting as Robert Stephenson proposed 278
tender, carried locomotive's fuel and water 12
Thirlwall, John, builds *My Lord* with Stephenson 118
Times, The
reporter carried away by his account of *Rocket*'s performance 197
reports *Cycloped* mishap 77
Titania, Robert Stephenson's yacht

311
account of life aboard 313
rough return passage from Norway 315
sails to Egypt 313
trip to Scotland 312–3
well supplied with the best cigars and wine 311
Tory party 231–2
Trade unions outlawed 111
Trafalgar, battle of 111
Trevithick 44
'could sell...all...engines' 51
a prospect in Wales 53
accounts never balanced 48
advantages of his engine 43
appeals to Parliament for compensation 272
appearance 43
army life unappealing 134
builds model steam carriage 45, 46
builds model steam engines 43
built first locomotive 15
closes *Catch-me-who-can* exhibition 58
conscripted into Bolivar's army 133
crocodilian attack 153
cuts losses after *Catch-me-who-can* exhibition closes 59
declines Blackett's invitation to build locomotive 59
deserts Bolivar's army 134
destitute in Colombia 153
develops high-pressure steam engine 43
dies in poverty, buried by his friends 273
discouraged by apathy to locomotives 59
early use of safety valve 53
embarks on South Seas whaler 95
engines at Science Museum 47
erects atmospheric engine 43
exhibits locomotive in London 57
final letter to Davies Gilbert 273–4
first boiler explosion 53
first trip to London 51
granted Peruvian mineral rights 132
Homfray association ends 57
human flight, iron ships and a gun carriage 271
inept businessman 48
influenced by Murdock's work? 47
influences Robert Stephenson in Colombia? 155
last South American adventure 152, 153
leaves England 95
leaves Peru for Costa Rica 134
legal dealings with Boulton & Watt 43
lionised in Lima 95
London carriage pelted with garbage 52
marriage 43
mines Peruvian copper 132
new high-pressure engine 51
obtains patent 51
Pen-y-darren, first locomotive in world 53
proposes a Reform memorial 273
reintroduced to his son 270, 271
romantic dreamer 95
safe passage home 153
salvages Peruvian shipwreck 131
sells model steam engine to London shop 59
sells portable engines 47
sells share in patent 53
simple traction experiments with carriage 48
steam carriage destroyed by fire 50
supplied drawings for Gateshead locomotive 57
supplies engine for boring cannon 53
swims out to *Esmeralda* to warn Cochrane 134
unsung hero 15
used feed pump 70
used fusible plug 212
Trevithick – *cont*

Uville changes his life 61
Viceroy for Peru grants passport 132
visits Coalbrookdale 51
vital dates 15
works on second road carriage 51
Tring, site of controversial chalk-cutting 277
try cock 163, 208
turnbuckles fitted to *Rocket*'s boiler stays 186
Turnbull, face permanently speckled 2
turnpike roads, faster than canals 126

Uville, Don Francisco
buys Trevithick's model from London shop 60
cash-flow problems 95
changes Trevithick's life 61
death 132
dishonest man 96
hopes to purchase engines in England 60
jealous of Trevithick 131
meets Captain Teague 61
Peruvian partnership agreement 96
returns to South America with Cornishmen 95
sails to Cornwall 59
second trip to England 59
successfully uses Trevithick's model steam engine in Andes 61
takes Trevithick's model engine back to Peru 60
tells stories of Peruvian riches 60
treachery 131
undermines Trevithick 131

valves
coupled to eccentrics 19
reversing sequence 170
supply steam to cylinders 19
synchronised with wheels 19
Vaughan's patent air blower 194
viewer, skilled men 87
Vignoles, Charles
aboard *Novelty* during demonstration 198
disliked George Stephenson 78
explanation for George Stephenson's hostility 102
likely author of attack on George Stephenson 223
resigns after difficulties working with George Stephenson 147
testifies for Robert Stephenson at Dee bridge inquest 307
Vivian, Andrew, Trevithick's cousin and business partner 50, 57

wagonways 86–7
Wakefield, *Rocket*'s regular driver 20
Walbottle Colliery
George Stephenson employed as engineman 105
Hackworth offered job after leaving Wylam 95
testimonial for Hackworth 84
Walker, James, respected engineer
locomotive estimates for L & M Railway 6
praises *Royal George* 99
report on fixed & locomotive engines 5
Walker Ironworks 123
Walmsley, Sir Joshua, accompanies George Stephenson to Spain 291
water jacket 182
Waters, Thomas, builds locomotive for Blackett 90–1
Watt, James, improved Newcomen's engine 41–2
welding 80–1, 327
Wellington Harmonic Band 246
Wellington, Duke of
abhorred locomotives 73
agrees to let ceremonies continue 254
concerns for his safety in the press of the crowd 258
dressed in mourning for King

George IV 247
kisses children 257
Napoleon's nemesis 73
national hero, despised Prime Minister 231
Prime Minister 73
special carriage 246
to open L & M Railway 244
Wheal Druid mine 48
wheels 33, 183
Wilkinson, John, eminent ironfounder 42
Winan's man-powered vehicle 29
Wood, Nicholas, respected engineer
accompanies George Stephenson to meet Pease 124
appointed judge 10
former viewer at Killingworth 87
introduces George Stephenson to Pease 125
met George Stephenson at sixteen 10
often accompanied George Stephenson to Wylam and Coxlodge 94
on good terms with George Stephenson 10
pessimistic prediction for locomotives 149
textbook author 10
worked at same colliery as George Stephenson 10
Wordsworth, John, poet's brother, commanded *Earl of Abergavenny* 72
workers, routinely exposed to risk 209
working class
agricultural labourers poorest paid 232
disposable income 73
lives not highly valued 7
reformers seek improvements in 131
Wortley, Stuart, later Lord Wharncliffe, one of the Grand Allies 94
wrought iron
available in only small quantities 81
details of manufacturing process 328–9
difficult to weld 81
malleable 82
more refined than cast iron 81
plates vary in thickness 82
quality variable 81
resistant to corrosion 80
stonger than cast iron 81
strengthened by beating 82
Wylam Colliery
concerns for jobs by horse handlers with locomotive success 92
employed Hackworth's father 85
first locomotive unsuccessful 91
George Stephenson regular visitor 94
locomotives set standard 94
owned by Blackett 57
Puffing Billy 92
second locomotive a great improvement 92
second locomotive described 93
second locomotive with return-flue boiler 91
traction experiments 88

yaw 206
yoke, of eccentric assembly 172, 173
Young, Robert, Hackworth's grandson
blames cracked casting for *Sans Pareil*'s demise 227
claims *Planet* was modelled on *Globe* 300

Credits

P. 3. Smiles, *Life of George and of his son Robert Stephenson, comprising also a History of the Invention and Introduction of the Railway Locomotive* (New York: Harper, 1868); pp. 11, 13. *Arcana of Science and Art*, 1830; p. 16 J. C. Jeaffreson, *The Life of Robert Stephenson, FRS* I (London: Longmans, 1866); p. 17 *Arcana of Science and Art,* 1830; p. 23 Smiles, *Life of George and Robert Stephenson*; p. 28 E. Galloway, *History and Progress of the Steam Engine; with a Practical Investigation of its Structure and Application* (London: Thomas Kelly, 183 6); p. 38 Worcester College, Oxford University; p. 44 F. Trevithick, *Life of Richard Trevithick, with an Account of his Inventions* I (London: Spon, 1872); p. 46 Trevithick, *Life of Richard Trevithick*; p. 52 Trevithick, *Life of Richard Trevithick*; p. 55 Smiles, *Life of George and Robert Stephenson*; p. 5 8 Trevithick, *Life of Richard Trevithick* I; p. 63 Science Museum/Science and Society Picture Library; p. 64 Science Museum/Science and Society Picture Library; p. 67 *Mechanics' Magazine*, 24 October 1829; p. 74 Church, *The Life of John Ericsson* I; p. 83 with permission from Jane Hackworth-Young; p. 89 Smiles, *Life of George and Robert Stephenson*; p. 98 R. L. Galloway, *The Steam Engine and its Inventors: A Historical Sketch* (London: Macmillan, 188 1); p. 101 Smiles, *Life of George and Robert Stephenson*, engraved by W. Hall after the portrait by John Lucas; p. 121 *The Engineer*, 21 August 1857; p. 126 Science Museum/Science and Society Picture Library; p. 139 Smiles, *Life of George and Robert Stephenson*; p. 166 drawn for author by Julian Mulock; p. 172 drawn for author by Julian Mulock; p. 203 Science Museum/Science and Society Picture Library; p. 224 Science Museum/Science and Society Picture Library; p. 237 National Portrait Gallery; p. 239 Smiles, *Life of George and Robert Stephenson*; p. 243 Smiles, *Life of George and Robert Stephenson*; p. 248 Manchester Public Library; p. 251 Science Museum/Science and Society Picture Library; p. 264 Smiles, *Life of George and Robert Stephenson*; p. 286 Science Museum/Science and Society Picture Library; p. 305 Smiles, *Life of George and Robert Stephenson.*